19th International Conference on Mechanical, Materials and Manufacturing Engineering (ICMMME 2017)

Venice, Italy
14 - 15 August 2017

ISBN: 978-1-5108-4663-0

Printed from e-media with permission by:

Curran Associates, Inc.
57 Morehouse Lane
Red Hook, NY 12571

Some format issues inherent in the e-media version may also appear in this print version.

World Academy of Science, Engineering, and Technology (WASET)
http://www.waset.org

Printed by Curran Associates, Inc. (2017)

Additional copies of this publication are available from:

Curran Associates, Inc.
57 Morehouse Lane
Red Hook, NY 12571 USA
Phone: 845-758-0400
Fax: 845-758-2633
Email: curran@proceedings.com
Web: www.proceedings.com

TABLE OF CONTENTS

Project and Experiment-Based Fluid Dynamics Education

Etsuo Morishita

***Abstract*—**This paper presents the project and experiment-based fluid dynamics education in Meisei University, a private institution in Tokyo, Japan. We pay attention not only to the basic engineering courses but also to the practical aspect of engineering experience. So, we prepare courses called the Projects from I to VI. The Projects I and II are designed for the first year, III and IV are designated for the second year, V and VI are prepared for the third year, respectively. Each supervisor is responsible for two of these projects every year. When students take the Project V and VI at the third year, we automatically assume that these students will join the lab of the project for the graduation thesis. We would like to show our experience in the Project I in the summer term, 2016. In this project, we introduce a traction flight vehicle called Cat Flyer. This is a kind of a kite towed by a car for example. This is very similar to parasailing, but flight is possible even on the roads. Experiments in mechanical engineering education are also very important, and we would like to explain our course on centrifugal pump, venture, and orifice. Although these are described in detail in the text books of fluid dynamics, it is still crucial to have practical experiments as a student.

***Keywords*—**Aerodynamics, experiment, fluid dynamics, project

I. Introduction

THIS paper describes the experimental and project-based fluid dynamics education in Meisei university. Fluid dynamics is sometimes a very difficult subject for many engineering students especially when the mathematical formulations are heavily involved. In Meisei university, we pay attention to the hands-on experience during the academic courses. It is interesting to note that major breakthrough is often brought by very practical people who are not necessarily academically excelled. One story is that bachelors contribute much more than people with doctor degrees in a huge manufacturing company in Japan. This is because that the practical experience in the real world is more useful as far as manufacturing is concerned. We are not saying that graduate education is not fruitful, but it depends on the situation. From this point of view, we weigh the project-based learning and the experiments in our courses.

II. Traction Flight Vehicle in Meisei Project I

A. Project I in the First Term 2016

We have about 400 new undergraduate students in engineering and science, 2016. There are altogether 34 subjects for these newcomers: 5 for physics, 6 for life sciences and biochemistry, 6 for mechanical engineering, 8 for electronics and electrical engineering, 5 for architecture, and 4 for ecology. There are 15 weeks to complete the project. First week was an orientation for all students. Each department also gives instructions for their own students.

B. Project I in Mechanical Engineering

We provide: Traction Flight Vehicle, Material Strength, Air Engine, Mini Model Car, Pet Bottle Rocket, and Shape Memory Alloy. One subject accepts about ten students and weekly seminars follow. We have altogether fifteen weeks to complete the project.

C. Traction Flight Vehicle in Project I

The traction flight vehicle is similar to parasailing [1]. Although there exists already flying taxis and bikes, they are very difficult to develop and might be dangerous in daily life circumstances. So, we propose a traction flight vehicle where the driver and/or the pilot might be on the flying part [2]. We allow the traction flight vehicle connected to the driving vehicle primarily with pitching motion as in Fig. 1 where U is driving velocity and ω is the angular velocity of pitching motion. We can achieve a stable traction flight by restricting the degrees of freedom of the towed wing.

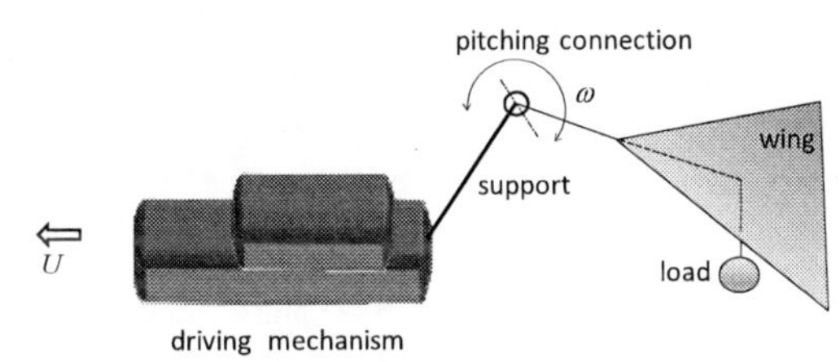

Fig. 1 Traction flight vehicle

After the first orientation, we explained the traction flight vehicle project. We made paper gliders [3] and model ground-effect vehicles [4] to get hands-on experience.

In the project, the lectures were given on the aerospace engineering and the aerodynamics which is not necessarily familiar in the mechanical engineering department. In many cases, airfoil theory and gas dynamics are missing in the mechanical syllabus. Then laboratory tour was conducted. We introduced several low-speed wind tunnels, water channel, and pump, venturi, orifice and weir. So, the students got the idea what about fluid dynamics and aerodynamics.

After the introductory process, the model parts of the traction flight vehicle were prepared.

Project I is also a course for the university open campus to the future engineers and scientists. Elementary school pupils

Etsuo Morishita is with Meisei University, Hodokubo, Hino, Tokyo, Japan (e-mail: etsuo.morishita@ meisei-u.ac.jp).

and the junior high school students were invited to show how to build and operate the traction flight vehicle. During the course of development, the author and the students both realized that the aerodynamic concept of the traction flight vehicle had to be explained to the participants. The students studied the ground effect by conducting a wind tunnel test for that purpose.

The students prepared the model traction flight vehicle and practiced how to work them. In the open campus, we had one session in the morning and two extra sessions in the afternoon. Each session accepted about ten students and altogether more than thirty pupils joined the project. We introduced the model traction flight vehicle, and operated the traction flight vehicle.

After the practice, the participants reviewed the project. They experienced the application of aerodynamics by the traction flight vehicle. One of the critical comments by the accompanying parents was that we should include the assembling process of the traction flight vehicle to understand the structural aspect of the model.

III. Fluid Dynamics Experiment

We provide a course called mechanical engineering experiment I for the summer term from April to July. Third year students are divided into four groups, so we have about ten students in each group. Instructors prepare also four experimental items, i.e., gel material deformation, machining, pumps, and heat transfer. The author is responsible for the pump experiment. Each categorized experiment continues for successive three weeks. For the first two weeks, students work on the experiment. The final third week, they complete the experimental report with tables, graphs, and some discussions. We introduce our fluid dynamics experiment, i.e., the pump experiment, in the following section.

A. Pump and Weir

Students first set the 2.2 kW two-pole pump discharge pressure by regulating a valve and flow rate is determined. Mass flow rate is measured directly by a classical balance. Students measure the time to fill the tank for a given mass. This becomes the reference mass and the volumetric flow rate Q. We also measure the water surface level of a regular triangular weir of 320 mm side length, Fig. 2, and obtain the volumetric flow rate Q form the hydrodynamic formula of JIS (Japan Industrial Standard). There are small discrepancies between the balance and the weir formula as shown in Fig. 3, where H is the weir head, K is a constant by JIS formula, and symbols denote students' group. But the agreement between the two methods is satisfactory from the practical point of view.

The pump suction pressure is measured and we can calculate the theoretical pump power. The pump motor input power is also measured, and we obtained the motor output power and its efficiency from a manufacturer's data chart. So, we finally get the efficiency of the pump η. Students draw the pump H(head) - Q curve, and the corresponding non-dimensional $\psi - \phi$ curve including the efficiency η as shown in Fig. 4, where $\phi \left(\equiv Q / nD^3\right)$ is the flow coefficient and $\psi \left(\equiv gH / n^2 D^2\right)$ is the head coefficient (D: diameter, g: gravitational acceleration, n: revolution per second). We also calculate the specific speed $n_s \left(\equiv \phi^{1/2} / \psi^{3/4}\right)$.

Fig. 2 Regular triangular weir of 320 mm side length

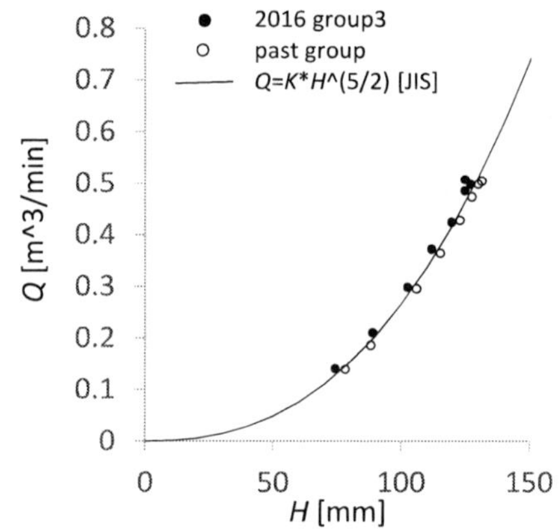

Fig. 3 Flow rate Q from triangular weir head H

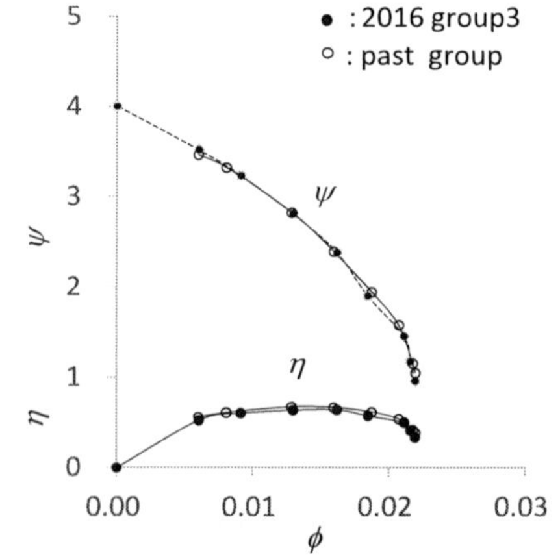

Fig. 4 Pump head coefficient ψ and efficiency η

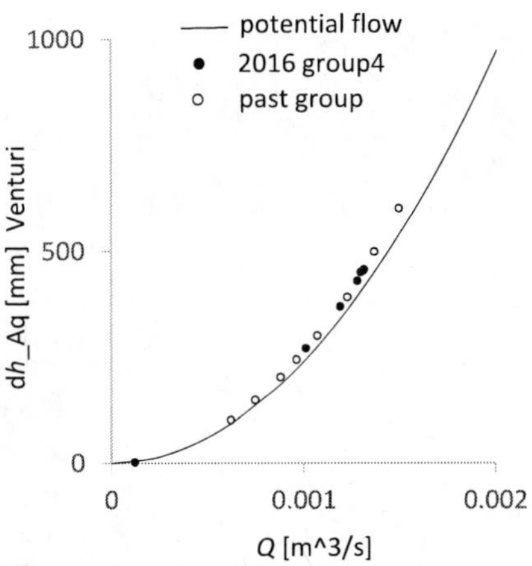

Fig. 5 Venturi pressure drop $\frac{d_2}{d_1}=\frac{23}{35}\approx 0.66 \quad [d]=\text{mm}$

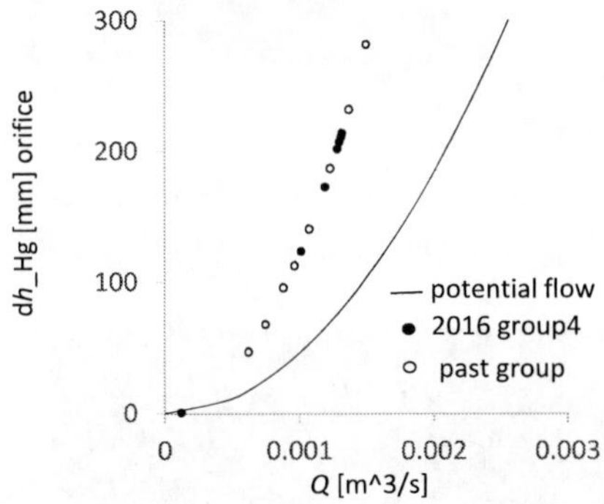

Fig. 6 Orifice pressure drop $\frac{d_2}{d_1}=\frac{19}{35}\approx 0.54 \quad [d]=\text{mm}$

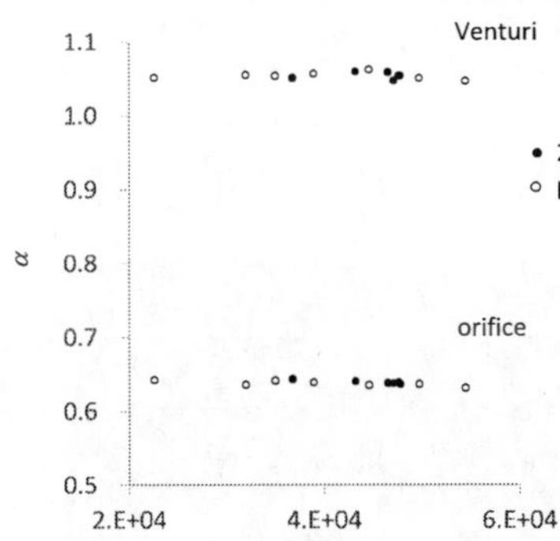

Fig. 7 Flow coefficient α

B. Venturi Tube and Orifice

One of the most famous fluid dynamic devices is a venturi. Students measure the pressure drop $\Delta h[\text{mm}](\Delta p[Pa])$ by a U-tube water manometer and realize that a flow through the venturi is nearly a potential flow with small viscous effect. The flow rate from the venturi pressure drop Δh is several percentage smaller than that by the Bernoulli's prediction as shown in Fig. 5. The inlet pipe inner diameter is d_1(=35 mm) and the throat diameter is d_2(=23 mm) .

An orifice is set in series behind the venturi and students experience higher pressure drop via the orifice than that of the venturi as shown in Fig. 6. Pressure drop is measured with a mercury-water U-tube manometer. The contraction is also higher for the orifice, with a throat diameter d_2=19 mm and a pipe diameter d_1=35 mm. Students have to pay attention not only to the density of mercury but also to that of water to calculate the pressure difference.

By changing the flow rate, we get the flow coefficient α of the ventrui and the orifice against the Reynolds number. The flow coefficient α is defined as the throat velocity divided by $\sqrt{2\Delta p/\rho}$, where ρ is the water density. These coefficients are nearly constant for the Reynolds number range tested as shown in Fig. 7.

IV. Visualized Potential Flow for Students

While our students conduct the fluid mechanics experiments, it might be very helpful if we can show the flow in the devices.

Although a complete viscous solution is possible these days, it is still very useful to show a simple potential flow image for basic understandings.

As for a two-dimensional nozzle which represents the venturi, we come to an idea that Laplace's equation may have analytical solution for the nozzle flow. Laplace's equation becomes

$$\frac{\partial^2\psi}{\partial x^2}+\frac{\partial^2\psi}{\partial y^2}=0 \tag{1}$$

where x, y is the coordinates, and ψ is the stream function. We assume the following form:

$$\psi=f(x)g(y) \tag{2}$$

The solution becomes as:

$$\frac{f''(x)}{f(x)}=-\frac{g''(y)}{g(y)}=c$$

$$f(x)=A\cosh\sqrt{c}\,x+B\sinh\sqrt{c}\,x \tag{3}$$

$$g(y)=C\cos\sqrt{c}\,y+D\sin\sqrt{c}\,y \tag{4}$$

where A, B, C, c and D are all constants. We may put

$$\psi(x,y)=e^{-x}\sin y \tag{5}$$

$$u = \frac{\partial \psi}{\partial y} = e^{-x} \cos y \quad (6)$$

$$v = -\frac{\partial \psi}{\partial x} = e^{-x} \sin y \quad (7)$$

$$p \equiv -u^2 - v^2 = -e^{-2x} \quad (8)$$

where p is the static pressure, u and v are the velocity components in the x - and y -directions, respectively. Figs. 8 and 9 show ψ and p, respectively. We can regard the constant stream function as a nozzle and/or a diffuser wall of the venturi. It must be emphasized to the students that the flow goes through the adverse pressure gradient as shown in Fig. 9. This situation causes problems like separation in the practical situation.

A real venturi is a pipe and therefore an axisymmetric solution is appropriate. Laplace's equation in this case becomes

$$\frac{\partial^2 \psi}{\partial r^2} - \frac{1}{r}\frac{\partial \psi}{\partial r} + \frac{\partial^2 \psi}{\partial z^2} = 0 \quad (9)$$

where r is the radial coordinate and z is the axial coordinate, respectively. The same separation of variables as in the two-dimensional case becomes as.

$$\psi = f(r)g(z) \quad (10)$$

$$\frac{g''}{g} = -\frac{f''}{f} + \frac{1}{r}\frac{f'}{f} = c \quad (11)$$

Solutions are obtained by the software Mathematica

$$f(r) = ArJ_1\left(\sqrt{c}r\right) + BrY_1\left(\sqrt{c}r\right) \quad (12)$$

$$g(z) = Ce^{\sqrt{c}z} + De^{-\sqrt{c}z} \quad (13)$$

where A, B, C, c and D are all constants, J and Y are the Bessel functions of the first and the second kind, respectively. One possible solution is

$$\psi = rJ_1(r)e^{-z} \quad (14)$$

The velocity components u_r, u_z and the pressure p become as:

$$u_z = \frac{\partial \psi}{r\partial r} = \left[\frac{J_1(r)}{r} + \frac{1}{2}\{J_0(r) - J_2(r)\}\right]e^{-z} \quad (15)$$

$$u_r = -\frac{\partial \psi}{r\partial z} = J_1(r) \cdot e^{-z} \quad (16)$$

$$p \equiv -u_r^2 - v_z^2 \quad (17)$$

Fig. 8 Two-dimensional nozzle theoretical streamline $\psi(x, y) = e^{-x} \sin y \quad (0 \le x \le 0.34, 0 \le |y| \le 1)$

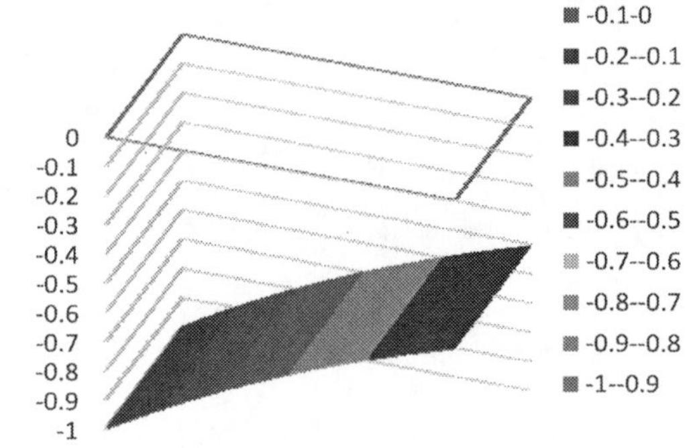

Fig. 9 Two-dimensional nozzle theoretical pressure p $(0 \le x \le 0.34, 0 \le |y| \le 1)$

Fig. 10 Axisymmetric nozzle theoretical streamline $\psi = rJ_1(r)e^{-z} \quad (0 \le z \le 1.3, 0 \le r \le 2)$

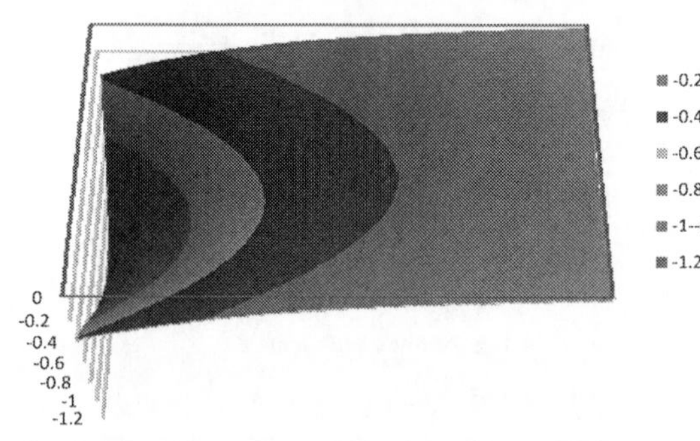

Fig. 11 Axisymmetric nozzle theoretical pressure p $(0 \le z \le 1.3, 0 \le r \le 2)$

Fig. 10 shows the stream function ψ , and Fig. 11 shows the pressure p for the axisymmetric case, respectively. These two-dimensional and axisymmetric solutions are mathematically periodic in the y- and the r-directions, respectively. The flow directions are reversible in the solutions, i.e., both diffuser and nozzle flows are possible.

Fig. 12 shows a potential flow stream function through a fictitious axisymmetric orifice with a contraction ratio about 2:1.

The finite difference from of (9) is solved numerically by spreadsheet [5]-[8]. Corresponding pressure is shown in Fig. 13. Real flow is viscous and separates, and also becomes turbulent and unsteady contrary to Figs. 12 and 13.

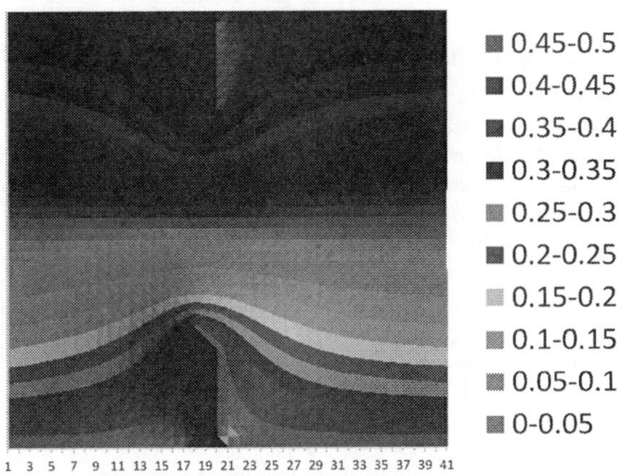

Fig. 12 Orifice potential flow stream function ψ
$(0 \le z \le 2, 0 \le r \le 1)$

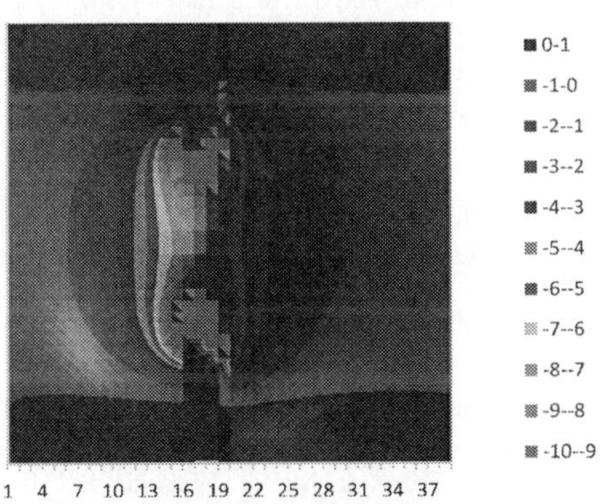

Fig. 13 Orifice potential flow pressure distribution p (outer boundary not included) $(0 \le z \le 2, 0 \le r \le 1)$

A flow through centrifugal pump impeller is visualized by the vortex singularity method in Fig. 14. The complex potential for the centrifugal pump impeller becomes as [9]-[11].

$$w = \frac{q}{2\pi}\ln z - i\frac{\Gamma_0}{2\pi}\ln z - \sum_{k=1}^{N}\sum_{j=1}^{n+1} i\frac{\Gamma_j}{2\pi}\ln\left(z - z_{k,j}\right) + i\frac{1}{2}r^2\omega \quad (18)$$

where N is the number of impellers, n+1 is the number of vortices on each impeller, q is the source strength, i.e., the pump flow rate, $z(= x + iy)$ is the complex number, $z_{k,j}$ is the coordinate of each vortex, Γ_0 is the inlet swirl, Γ_j is the circulation of each vortex, and ω is the angular velocity.

The imaginary part of (18) is the stream function as:

$$\psi = \frac{q}{2\pi}\theta - \frac{\Gamma_0}{2\pi}\ln r - \sum_{k=1}^{N}\sum_{j=1}^{n+1}\frac{\Gamma_j}{2\pi}\ln\left|z - z_{k,j}\right| + \frac{1}{2}r^2\omega \quad (19)$$

where θ is the circumferential coordinate. From (19):

$$\begin{aligned} w_r &= \frac{\partial\psi}{r\partial\theta} \\ &= \frac{q}{2\pi r} - \sum_{k=1}^{N}\sum_{j=1}^{n+1}\frac{\Gamma_j}{2\pi}\left[\frac{-\left(x - x_{k,j}\right)\sin\theta + \left(y - y_{k,j}\right)\cos\theta}{\left(x - x_{k,j}\right)^2 + \left(y - y_{k,j}\right)^2}\right] \end{aligned} \quad (20)$$

$$\begin{aligned} w_\theta &= -\frac{\partial\psi}{\partial r} \\ &= \frac{\Gamma_0}{2\pi r} + \sum_{k=1}^{N}\sum_{j=1}^{n+1}\frac{\Gamma_j}{2\pi}\frac{\left(x - x_{k,j}\right)\cos\theta + \left(y - y_{k,j}\right)\sin\theta}{\left(x - x_{k,j}\right)^2 + \left(y - y_{k,j}\right)^2} - r\omega \end{aligned} \quad (21)$$

where w_r and w_θ are the radial and circumferential relative velocity components, respectively, and

$$\cos\theta \equiv \frac{\partial x}{\partial r} = \frac{\partial y}{r\partial\theta}$$
$$\sin\theta \equiv \frac{\partial y}{\partial r} = -\frac{\partial x}{r\partial\theta}$$

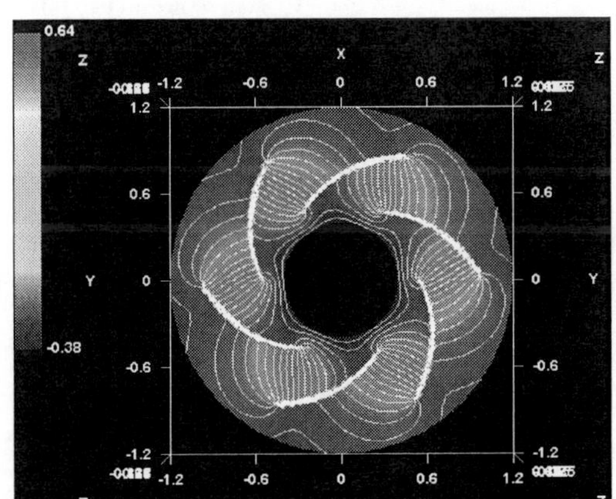

Fig. 14 Potential flow pressure p through centrifugal pump impeller
$N = 6,\ \beta = \pi/3,\ \phi = 0.24,\ r_1/r_2 = 0.54$

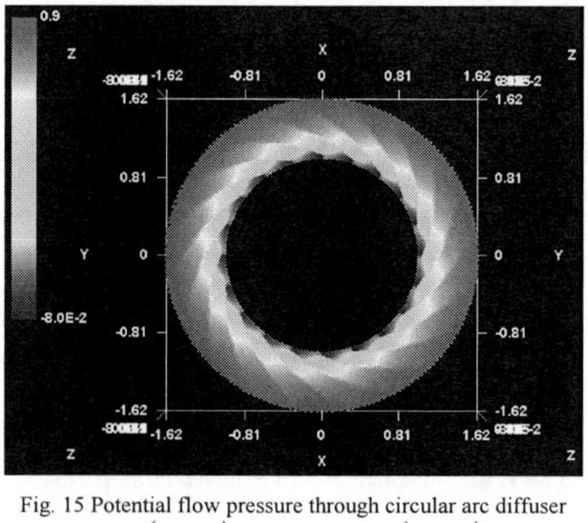

Fig. 15 Potential flow pressure through circular arc diffuser $N = 20$, $1.04(= r_3/r_2) \le r_{\text{diffuser}}/r_2 \le 1.59(= r_4/r_2) = 1.59$, $\alpha = 20^\circ$

A logarithmic spiral for the impeller with the inclination β is given by

$$r = e^{-\tan\beta \cdot \theta} \tag{22}$$

The flow tangency condition on the impeller becomes

$$\frac{dr}{rd\theta} = \frac{w_r}{w_\theta} = -\tan\beta \tag{23}$$

From (20), (21) and (23), we get a non-dimensional form of (23) on the impeller collocation point (x_{ci}, y_{ci}) as:

$$\sum_{k=1}^{N}\sum_{j=1}^{n}\frac{\Gamma_j}{2\pi}\left\{\left[\frac{(x_{ci}-x_{k,j})(\tan\beta\cos\theta+\sin\theta)+(y_{ci}-y_{k,j})(\tan\beta\sin\theta-\cos\theta)}{(x_{ci}-x_{k,j})^2+(y_{ci}-y_{k,j})^2}\right]\right\} \tag{24}$$

$$= \left(r_{ci} - \frac{\Gamma_0}{2\pi r}\right)\tan\beta - \frac{\phi}{r_{ci}} \quad (1 \le i \le n)$$

where

$$x_{ci} \equiv \frac{x_{1,j}+x_{1,j+1}}{2}, \quad y_{ci} \equiv \frac{y_{1j}+y_{1,j+1}}{2} \tag{25}$$

$$r_{ci} \equiv \sqrt{x_{ci}^2 + y_{ci}^2} \tag{26}$$

$$\phi \equiv \frac{q}{2\pi r_2 \cdot (r_2\omega)} \tag{27}$$

Although we use the same symbols as those of (18) ~ (23) in (24) ~ (26), the coordinates are divided by the impeller outer radius r_2, the velocities are divided by the impeller tip speed $u_2 = r_2\omega$, and the circulations are divided by $u_2 r_2$. In (24), we use the Kutta condition as:

$$\Gamma_{n+1} = 0 \tag{28}$$

Therefore, the summation by j ends at n in (24).

We can solve (24) to get $\Gamma_j (1 \le j \le n)$, and therefore determine the velocity components form (20) and (21). Bernoulli's equation for pressure becomes

$$\frac{1}{2}(w_r^2 + w_\theta^2) + p - \frac{1}{2}r^2 = \text{const.} \tag{29}$$

where the equation is non-dimensional divided by ρu_2^2.

If we omit the last term of the right-hand-side of (18), the equation becomes the complex potential for a stationary diffuser. Similar simultaneous equations are obtained from the flow tangency conditions on the diffuser blades as (24). We can get pressure from Bernoulli's equation in the stationary coordinate. Pressure recovery through circular arc diffuser vanes is shown in Fig. 15. The diffuser vane inclination at the inlet radius r_3 is about 20 degree. Inlet flow angle at the impeller outer radius r_2 is also set $\alpha = 20$ degree, while the circumferential velocity component at the impeller exit is u_2. The diffuser vane geometric parameters are given in [12].

V. Conclusion

We introduce the project- and experiment-based fluid dynamics education in Meisei University, Tokyo, Japan. We proposes a traction flight vehicle. This concept is used in the project for the university students and the open campus for nearby pupils. Students get hands-on experience to build the traction flight vehicle and some teaching skills to elementary and junior high school pupils. We test centrifugal pump, triangular weir, venturi, and orifice in the fluid dynamics experiment. Although we learn these devices in the lectures, it is worthwhile for the mechanical engineering students to see how they work in the practical situation.

Acknowledgment

The present author would like to express many tanks to the Project I students and the mechanical engineering experiment students. Special thanks to Mr. H. Matsuda and TA Mr. H. Watanabe.

References

[1] E. Morishita and T. Shibuya, "Traction flight vehicle," JP2014196046 (A) 2014-10-16.
[2] T. Shibuya, T. Ishido and E. Morishita, "Cat Flyer—A Novel Wing in Ground-Effect Vehicle (CD)," APISAT2013, Takamatsu, Japan, Nov. 21, 2013.
[3] http://www.sciencetoymaker.org/airsurf/images/GLIDERS/MakeMosquitoGliderPDF.pdf (accessed on 19/12/2016)

[4] http://www.instructables.com/id/Paper-ground-effect-vehicle-2/ (accessed on 19/12/2016)
[5] C. Misner and P. Cooney, Spreadsheet Physics, Addison-Wesley, Reading, MA (1991), pp.43-56.
[6] J. Walkenbach, Excel 97 Bible, IDG Books, Foster City, CA (1996), pp.171-237.
[7] K.A. Hoffmann and S.T. Chiang, Computational Fluid Dynamics for Engineers, Vol.1, Engineering Education System, Wichita, KS (1995), pp.288-343.
[8] E. Morishita. "Spreadsheet Fluid Dynamics," Journal of Aircraft, Vol. 36, No. 4 (1999), pp. 720-723.
[9] A. J. Acosta, "An Experimental and Theoretical Investigation of a Two Dimensional Centrifugal Pump Impeller," Transaction of ASME, Vol. 76. (1954), pp.749-763.
[10] M. Mizutani, G. Kamimoto and H. Mizutani, "Volute Pump Analysis by Singularity Method (in Japanese)", Turbomachinery, Vol.10, No.8 (1982) pp.459-466.
[11] N. Kaiwa, T. Ishido and E. Morishita, "Flow and Characteristic Analysis of Centrifugal Pump by Vortex Method (in Japanese)," JSME Kanto Proc. 2012(18), pp.125-126.
[12] E. Morishita, Centrifugal compressor diffusers, M.Sc. Dissertation, University of Cambridge, 1983.

World Academy of Science, Engineering and Technology
International Journal of Mechanical, Aerospace, Industrial, Mechatronic and Manufacturing Engineering Vol:11, No:5, 2017

Optimization of the Aerodynamic Performances of an Unmanned Aerial Vehicle

Fares Senouci, Bachir Imine

***Abstract*—**This document provides numerical and experimental optimization of the aerodynamic performance of a drone equipped with three types of horizontal stabilizer. To build this optimal configuration, an experimental and numerical study was conducted on three parameters: the geometry of the stabilizer (horizontal form or reverse V form), the position of the horizontal stabilizer (up or down), and the landing gear position (closed or open). The results show that up-stabilizer position with respect to the horizontal plane of the fuselage provides better aerodynamic performance, and that the landing gear increases the lift in the zone of stability, that is to say where the flow is not separated.

***Keywords*—**Aerodynamics, wind tunnel, turbulence model, lift, drag.

I. Introduction

THE stability and aerodynamic performance of the aircraft are based on several parameters. Among these parameters, the stabilizer plays a major role on the maneuverability and the pitch and yaw control and the landing gear may present a brake on the advancement. Such importance is significant in several phases of flight (takeoff, landing, maneuvering, etc.) with well-known consequences: increased drag, partial or total loss of control of the aircraft [1], [2], etc.

This work presents a numerical and experimental study of a subsonic flow around a drone of four configurations: Unmanned Aerial Vehicle (UAV) with reverse V form stabilizer, drone with horizontal stabilizer up position, drone with horizontal stabilizer down position, and a drone with landing gear (Figs. 1-3) [3]. The first phase of this work is to build a prototype on SolidWorks to be exported to a mesh generator where a computational domain is discretized. A computer code is used to estimate the lift and drag coefficients at a speed of 20 m/s. The Spalart-Allmaras turbulence model is used to solve the Navier-Stokes equations [4]. The second phase consists of building a model of a 1/7-scale corresponding to the dimensions of the test vein of the wind tunnel. The tests in the wind tunnel are conducted at a 20 m/s speed for an angle of attack variation from 5° to 17° to determine the lift and drag coefficients. The wind tunnel tests and the numerical tests were effectuated and compared with each other. These experiments allowed us to bring out the main trends in order to obtain an optimal design and to validate a mathematical model.

Fares Senouci is with the Oran University of Science and Technology-Mohamed Boudiaf, Algeria (e-mail: senouci.fares41@gmail.com).

II. Description of UAVs

The first drone is shown in Fig. 1. It has a 0.64-m wingspan and swept wing of 3.5° a dihedral of 4° and an empennage of NACA 0012 in reverse V form, and a 0.035-m^2 area and a 0.65-m chord.

Fig. 1 UAV with landing gear

The second drone is shown in Fig. 2. It is a UAV without landing gear it has an up-horizontal stabilizer. It has a 0.64 m wingspan and swept wing of 3.5° a dihedral of 4° and an empennage of NACA 0012 in reverse V form, and a 0.035-m^2 area and a 0.65-m chord.

Fig. 2 UAV with an up-horizontal stabilizer and without landing gear

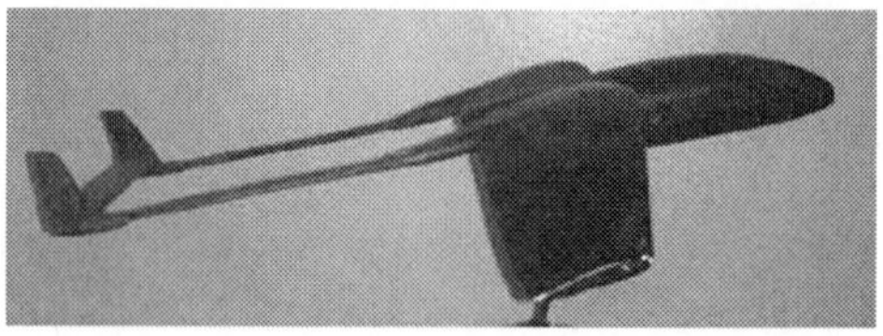

Fig. 3 UAV with a down-horizontal stabilizer and without landing gear

The third UAV is shown in Fig. 3. It is a UAV without landing gear and it has a down-horizontal stabilizer. It has a 0.64-m wingspan and swept wing of 3.5° a dihedral of 4° and an empennage of NACA 0012 in reverse V form, and a 0.035-m^2 area and a 0.65-m chord.

III. Transport Equation and Turbulence Model

The flow around the UAV is considered turbulent and symmetric. The general form of the transport equations can be written in Cartesian coordinates as follows:

The continuity equation in its general form:

$$\frac{\partial}{\partial xj}(\bar{\rho}\tilde{U}j) = 0 \quad (1)$$

And the conservation equation of the moment in its general form:

$$\frac{\partial}{\partial x_j}(\bar{\rho}\tilde{U}i\tilde{U}j) = \bar{\rho}\, g_i - \frac{\partial \bar{p}}{\partial x_i} - \frac{\partial}{\partial x_j}(\rho\tilde{u}_i\tilde{u}_j) + \frac{\partial}{\partial x_j}\left[\bar{\mu}\left(\frac{\partial \tilde{U}i}{\partial x_j} + \frac{\partial \tilde{U}j}{\partial x_i}\right) - \frac{2}{3}\bar{\mu}\frac{\partial}{\partial x_k}\delta ij\right] \quad (2)$$

where U is the flow velocity.

Turbulence Model [4]: The turbulence model used in this work is a model with a transport equation for the quantity suggested by Spalart and Allmaras.

$$\frac{\partial(\bar{\rho}\vartheta)}{\partial t} + \vec{\nabla}.(\bar{\rho}\,\vec{V}\hat{v}) = \vec{\nabla}.\left[\frac{\bar{\mu}+\bar{\rho}\,\tilde{v}}{\partial SA}\,\vec{V}\hat{v}\right] + \mathrm{P}_v - D_v \quad (3)$$

where the terms of production and destruction are defined as:

$$P_v = C_{b1}C_\mu\tilde{S}\bar{\rho}\tilde{v} + C_{b2}\frac{\bar{\rho}(\nabla\tilde{v})^2}{\partial SA} \quad (4)$$

$$D_v = C_{w1}f_w\frac{\bar{\rho}(\nabla\tilde{v})^2}{\rho d^2} \quad (5)$$

$$f_x = \mathrm{g}\left(\frac{1+C_{w3}^6}{\mathrm{g}^6+C_{w3}^6}\right) \quad (6)$$

with $\mathrm{g} = r + C_{w2}(r^2 - r)$ and

$$M_t = \bar{\rho}\tilde{v}\frac{x^3}{x^3+C_{v1}} \qquad x = \frac{\bar{\rho}\tilde{v}}{\mu} \quad (7)$$

The closing coefficients are given by the following values:

$$\sigma SA = \frac{2}{3}\,;\ C_{b1} = 0.1355\,;\ C_{b2} = 0.622 \quad (8)$$

$$C_{v1} = 7.1\,;\ C_{w1} = 0.3\,;\ C_{w3} = 2 \quad (9)$$

IV. Experimental Installation

The experiments were realized in a subsonic wind tunnel with closed circuit and 0.46 m × 0.46 m test section.

In the test section, the speed was fixed at 20 m/s (which corresponds to 2.10^5 Reynolds number).

The wind tunnel is horizontal type. A conventional closed-circuit wind tunnel has advantages over an open-circuit wind tunnel (Fig. 4); especially, a higher flow velocity, a lower energy consummation, and a lower noise level. The air velocity around the UAV is produced by an axial flow motor and produces a maximum velocity of 69 m/s [5]. In order to make tests in the wind tunnel, a prototype is drawn on SolidWorks to be able to produce a wooden model of a 1/7-scale and 0.425 m wingspan which corresponds to the dimensions of the test vein.

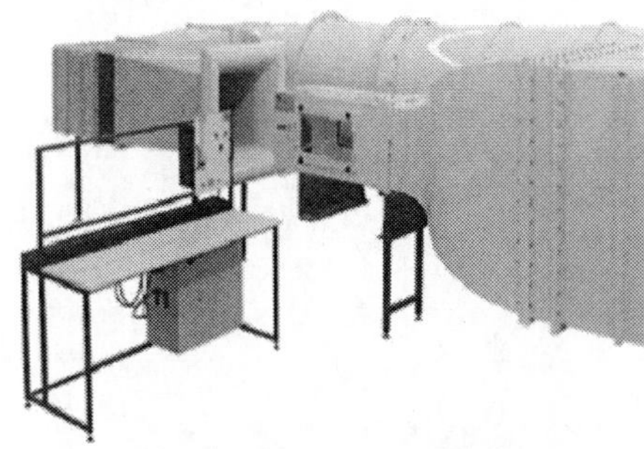

Fig. 4 Subsonic wind tunnel

Fig. 5 Model in the wind tunnel

V. Mesh and Boundary Conditions

Like most of the problems involved in fluid flow, a virtual wind tunnel is an ideal configuration. All three-dimensional problems could be divided along the longitudinal plane of symmetry [5].

After the models were imported from SolidWorks, a field was created around the UAV to analyze the flow. It is important to examine the asymmetries of the mesh elements to ensure that there are no highly asymmetric elements present in the mesh [5]. One should recall that very uneven elements can cause solution convergence problems in FLUENT [6].

The boundary conditions must be defined by GAMBIT. An initial speed condition at the input is used to define the flow velocity at the inlet. The front face of the flow control volume is imposed as an initial condition for the input speed. To specify the flow properties entering the volume or calculation domain in addition to the face before, the left, top, and bottom faces are also specified to ensure that the flow properties are taken into account by the FLUENT solver. The outflow plane is specified as an output condition of this flow. The surface of the UAV is specified as a boundary of the wall to differentiate the solid and liquid regions. Finally, the plane of symmetry is specified as a plane of symmetry. A summary of the boundary conditions used for this UAV is shown in Fig. 7.

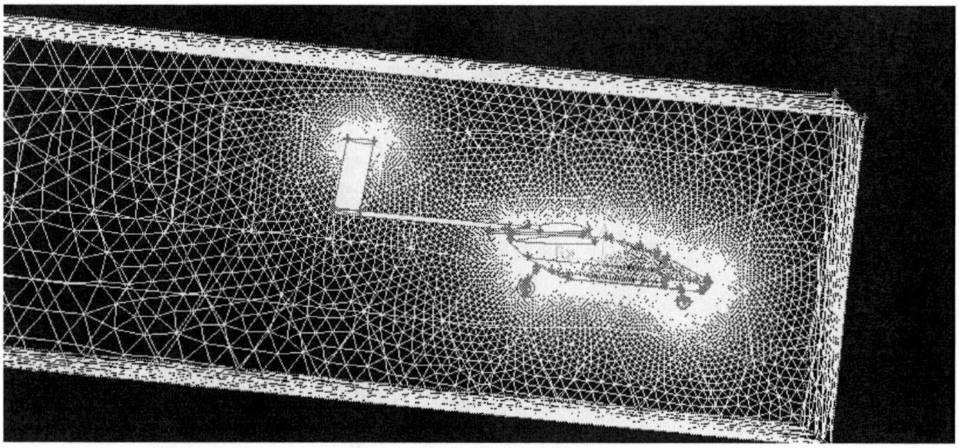

Mesh and boundary conditions

Fig. 6 The mesh landing gear UAV

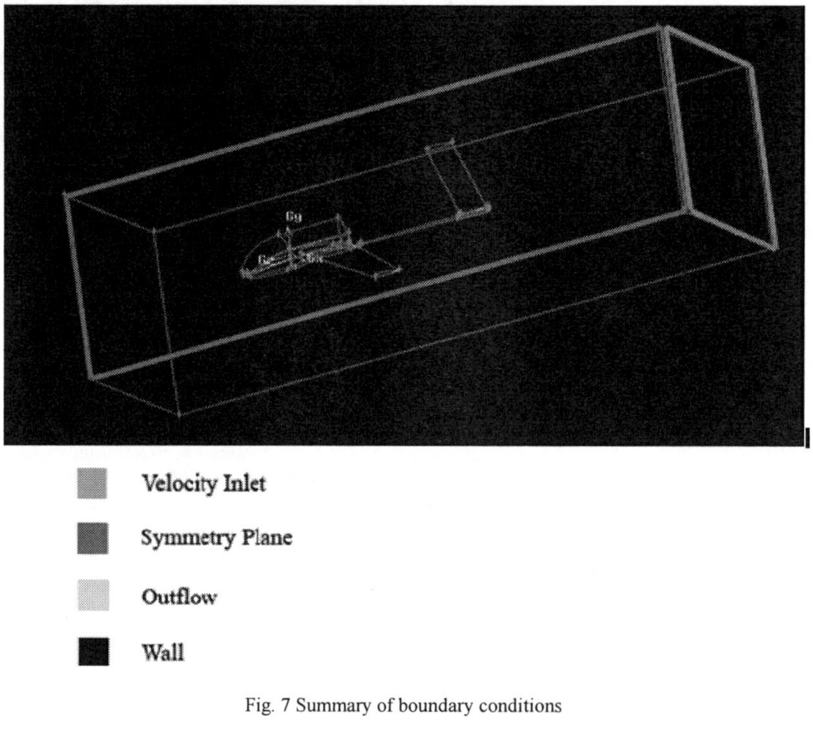

Fig. 7 Summary of boundary conditions

VI. Results and Discussions

Several results have been found for several angles of attack which varying between -5° and 17° to determine the lift and drag of the three types of UAVs for a 20-m/s air flow.

Fig. 8 compares the lift coefficients obtained numerically with that obtained experimentally for a reverse V form UAV with a landing gear. A difference of 24% is observed on the evolution of the lift in favor of the experiment and an increase of the point of stall of 4° in favor of the digital [7], it is certainly to the choice of the model of turbulence which remains imprecise in the separates flux zone.

The experimental lift coefficients comparison of UAV with

and without landing gear is shown in Fig. 9. The landing gear increases the lift in the zone where the flow is stable, while in the flow separation zone, the lift is not affected by the presence of the landing gear.

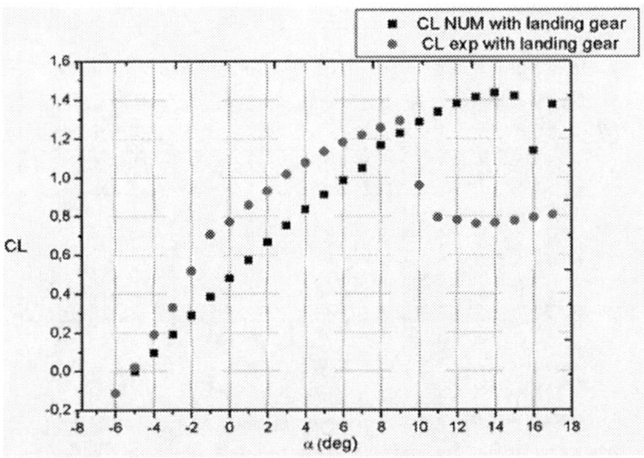

Fig. 8 Evolution of the numerical and experimental lift coefficients of the UAV with landing gear and reverse V form of stabilizer

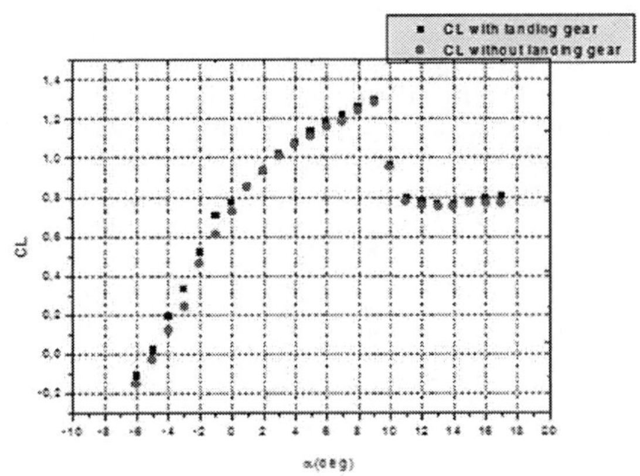

Fig. 9 Evolution of the experimental lift coefficients of the UAV with and without landing gear

The lift of the UAV with up-horizontal stabilizer is much better than the UAV with down-horizontal stabilizer. This is due to the presence of vortices generated by the fuselage and the wing which affect the down-horizontal stabilizer Fig 10.

If we compare two drones of the same category (UAV with up-horizontal stabilizer and UAV with reverse V form) as shown in Fig. 11, the up-horizontal stabilizer pushes the stall to 2 degrees.

Fig. 12 compares the drag coefficients obtained numerically with that obtained experimentally for a reverse V form of stabilizer of UAV with landing gear.

The gap is in favor of the UAV with a reversed V form of stabilizer throughout the evolution of the two curves

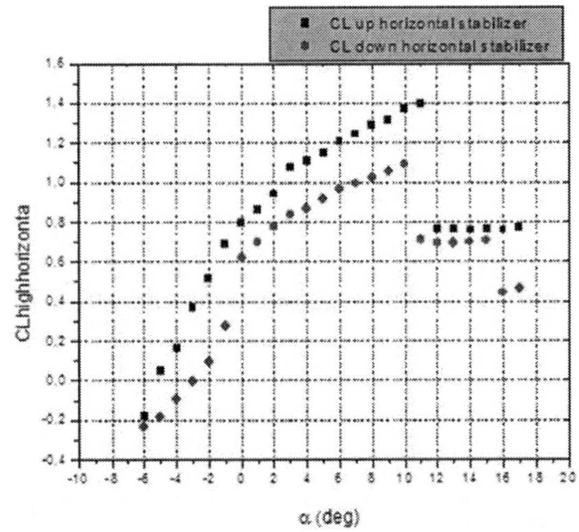

Fig. 10 Evolution of the experimental lift coefficients of the up and down-horizontal stabilizer

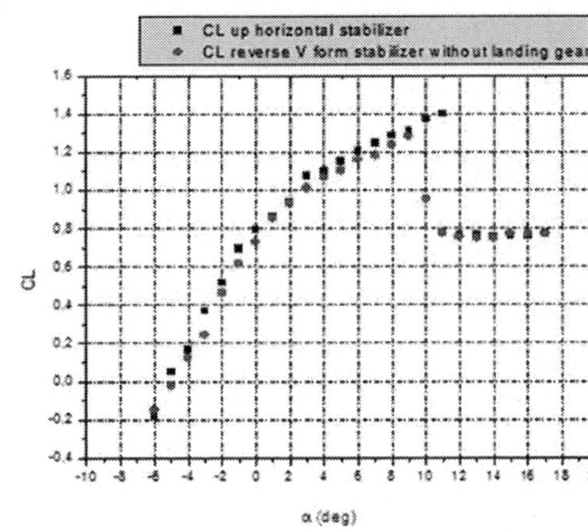

Fig. 11 Evolution of the experimental lift coefficients of the up-horizontal stabilizer and reverse V form of stabilizer

Fig. 13 represents the evolution of the experimental coefficients of drag as a function of the angle of attack of the UAV with up-horizontal stabilizer and UAV with down-horizontal stabilizer. It can be seen that the up-horizontal stabilizer has less drag than the down-horizontal stabilizer.

Fig. 14 represents the evolution of the experimental drag coefficients as a function of the angle of attack of the UAV with landing gear and of the UAV without landing gear. The landing gear represents a light braking on the advancement.

Fig. 15 compares the drag of the UAV with up-horizontal stabilizer and the UAV with reverse V stabilizer. Unlike the lift, the up-horizontal stabilizer breaks the progress of the drone compared to the UAV with reverse V stabilizer form.

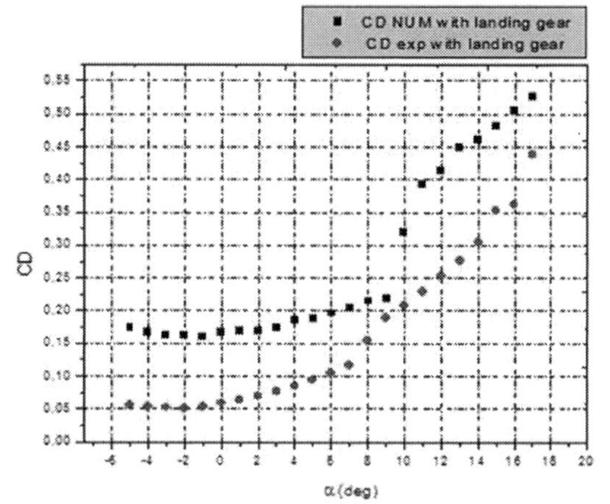

Fig. 12 Evolution of the experimental and numerical drag coefficients of the reverse V form of stabilizer for a UAV with landing gear

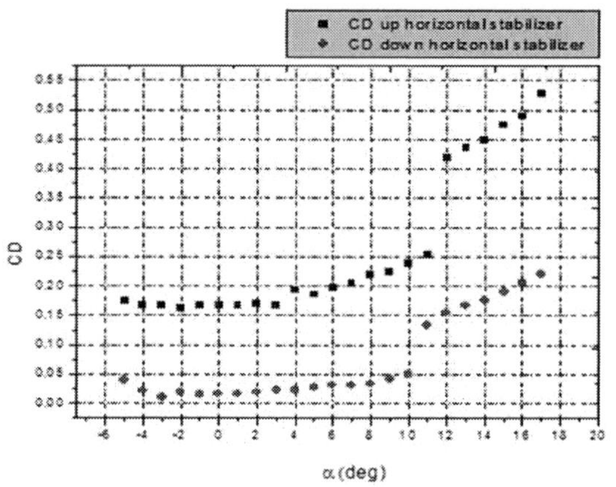

Fig. 13 Evolution of the experimental drag coefficients of up-horizontal stabilizer and down-horizontal stabilizer

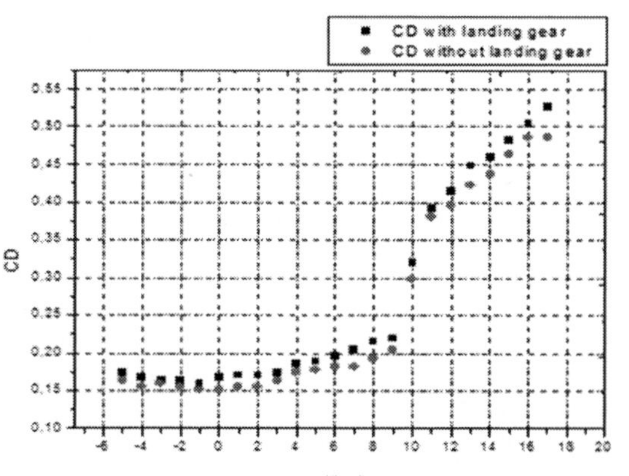

Fig. 14 Evolution of the experimental drag coefficients of reverse V form of stabilizer for an UAV with and without a landing gear

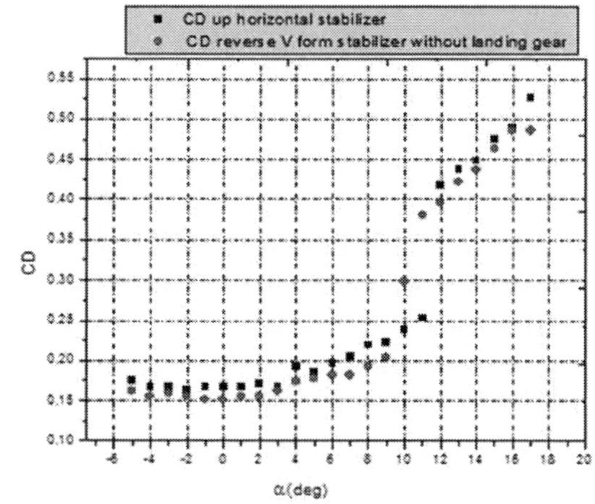

Fig. 15 Evolution of the experimental drag Coefficients of reverse V form of stabilizer for an UAV without landing gear and UAV with up-horizontal stabilizer

VII. Conclusion

This study examines the influence of stabilizer and the presence of the landing gear on the aerodynamic performance of UAVs. The study focuses on the optimization of the shape of the stabilizer (in reverse V or horizontal) and their position (up and down) and the landing gear (closed or open position). A numerical study was also conducted.

References

[1] O. M. Fouatih, et al.—Design optimization of the aerodynamic passive flow control on NACA 4415 airfoil using vortex generators‖, DOI: 10.1016/j.euromechflu, 2015.11.006.
[2] Elsa M. Cárdenas et al.—Design of an Unmanned Aerial Vehicle for Ecological Conservation‖ DOI: 10.2514/6.2005-7056·
[3] Luis Velazquez, —low-speed wind tunnel testing for an unmanned aerial vehicle airfoil‖, Institute of Thermomechanics AS CR, v.v.i., Prague, October 21 - 23, 2009.
[4] A. Zeghib, Revue des Energies Renouvelables CISM'08 Oum El Bouaghi, 311 – 324, 2009.
[5] Manuel de la TE 44 Subsonic Wind Tunnel.
[6] Fluent 62 User's Guide. Fluent, Inc, 2005.
[7] F. Senouci, B. Imine—Numerical and Experimental Study of Flow around an UAV ‖SKP Engineering college, TN, India, 16-17, 2011, pp. 400–404, December 2011, unpublished.

Open Jet Testing for Buoyant and Hybrid Buoyant Aerial Vehicles

A. U. Haque, W. Asrar, A. A. Omar, E. Sulaeman, J. S Mohamed Ali

***Abstract*—**Open jet testing is a valuable testing technique which provides the desired results with reasonable accuracy. It has been used in past for the airships and now has recently been applied for the hybrid ones, having more non-buoyant force coming from the wings, empennage and the fuselage. In the present review work, an effort has been done to review the challenges involved in open jet testing. In order to shed light on the application of this technique, the experimental results of two different configurations are presented. Although, the aerodynamic results of such vehicles are unique to its own design; however, it will provide a starting point for planning any future testing. Few important testing areas which need more attention are also highlighted. Most of the hybrid buoyant aerial vehicles are unconventional in shape and there experimental data is generated, which is unique to its own design.

***Keywords*—**Open jet testing, aerodynamics, hybrid buoyant aerial vehicles, airships.

I. Introduction

AIRSHIPS were the queen on the sky in early ages; where the major design and development of these flying machines was carried out during the period of first world war. England, France, Germany, Italy, and USA were the countries who contributed in such development programs. Applications for these vehicles include scientific data collection, communications relay, and transportation of goods and for the tourism industry. Wind tunnel testing plays a dominant role for the accurate prediction of aerodynamic and stability characteristics of buoyant vehicles. Open jet as well as wall bounded wind tunnel have been used in the past and are also being utilized presently for different buoyant and hybrid buoyant configuration; the same has resulted in many useful data on the aerodynamic and stability. Hybrid buoyant aerial vehicles *(HBAVs)* have been the focus of research. There is a significant growth potential of the *HBAVs*, the design and development phase of which will be requiring aerodynamic data with flow quality and data uncertainty. The cost spent on the wind tunnel testing of such vehicles is quite significant. Wind tunnels represent a useful tool for aerodynamic and stability study on different airships and hybrid buoyant aircraft. An increase in the number of wind tunnel tests performed on different configurations can be observed from the trend plot line shown in Fig. 1. In this figure, the authors have randomly arranged the decade wise (from 1910 to 2015) wind tunnel testing performed on buoyant and hybrid buoyant vehicles. Data is collected from different sources, hence no reference is provided here. No half model testing has been done as one cannot simulate the yaw angle for the scaled down models as the pitch and yaw axis of the body is asymmetric. Dynamics of *HBAVs* are distinct from those of heavier-than-air craft due to the added mass effects, which can be estimated by using towing tank.

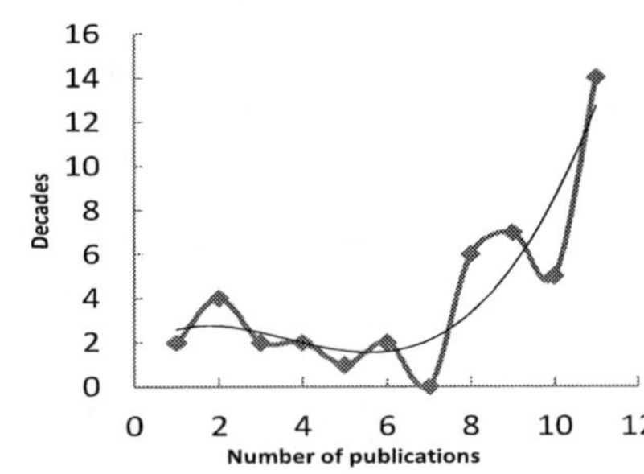

Fig. 1 Trendline plot of the decade wise wind tunnel testings

In the early decades, a limited number of experimental testings were performed on different scaled down models of airships. To the authors' best knowledge, the first model tested in the wind tunnel was of L-33 rigid airship in 1917 [1], followed by the testing of scaled down model of R-29 in 1920, [2]. Late, pressure distribution over the hull of R-101 was experimentally determined along with the estimation of the hinge moments [3]. These calculations, made at the request of the Zeppelin Airship Company of Friederichshafen, Germany, were based on the shape of the ZR III, with the following simplifications: cars, fins, and rudders removed; all cross sections replaced by equivalent circular cross sections. From 1991-2000, about seven different configurations were tested [4]–[10] and this number slightly decreased by two during 2001-2010, [11]–[15]. The highest number of publications were during 2011-2014, [16]–[29]. History of wind tunnel and airships are closely related to one another. Perhaps the first wind tunnel was manufactured but later more efforts were put towards aircraft [30]. Its design and construction was funded by German society for Airship study. It was a closed circuit wind tunnel with rectangular cross section [31].

On the analytical side, fundamental drag equation for

A. U. Haque is with the Experimental & Computational Thermofluid Mechanics Research Group (ECTMRG), Department of Mechanical Engineering. International Islamic University Malaysia (IIUM), 50728 Kuala Lumpur, Malaysia (e-mail: anwar.haque@live.iium.edu.my).

W. Asrar, E. Sulaeman and J. S. Mohamed Ali are with the Experimental & Computational Thermofluid Mechanics Research Group (ECTMRG), Department of Mechanical Engineering. International Islamic University Malaysia (IIUM), 50728 Kuala Lumpur, Malaysia (e-mail: waqar@iium.edu.my, esulaeman@iium.edu.my, jaffar@iium.edu.my).

A. A. Omar is Professor, Department of Aeronautical Engineering, University of Tripoli, Libya (e-mail: aao@aerodept.edu.ly).

aircraft was also derived from airships [32]. Also, Munk-Multhopp's work on pitching moment was also based on the experimental work on airships by Munk [33].

In the case of hybrid airships; the experimental data for aerodynamic are static stability derivatives is quite limited [32]. One thing common in the configurations of *HBAVs* and buoyant vehicles (airships) is the voluminous hull/fuselage which is filled by the lifting gas to provide aerostatic lift. Such configurations were tested in open jet/closed test section in the past, a few of which are shown below in Table I:

TABLE I
FEW WIND TUNNEL TESTINGS

S No.	Name	Reference	Type
1	Goodyear Zeppelin	[34]	Open jet+closed
2	Arkon	[9]	Open jet
3	LZ-120	[35]	Open jet
3	Zhiyuan-1	[11],[18]	Open jet
4	ESTOLAS	[36]	Open jet
5	Dynalifter	[37]	Closed
7	HAV	[38]	Closed
8	*IIUM-HBAV*	[39]	Closed

Most of the experimental data is related to the lifting bodies of hypersonic vehicles, tested inside the test section of the wind tunnel. For example, some experimental data in an old *NACA* report [40] which is based on $W\text{-}F_2$ configuration. As per Ash [41], this was a modified design of $M_2\text{-}F$ with modifications to the after body, the control surfaces, and the canopy location. Others include that on *NASA HL-20* lifting body concept [42]. All such bodies have λ greater than six and have wings blended with the fuselage. Recent interest in the configuration design of hybrid HBAVs with the hull of lifting profile has put a question about the limitations of the wind tunnel testing of such hybrid buoyant vehicles, inside the test section.

Before going into further discussion on the question in research, it is important to first define the nomenclature for the said aerial vehicles, so that a common terminology exists for all; either working in academia or in industry. We are perhaps not the first to highlight that the term hybrid is not fully explored, as the same has recently been pointed out by [43] and [32]. The term hybrid is applicable for hybrid airship but not for the airship which;

> *"refers to any air vehicle which depends on a significant amount of buoyant lift to obtain cruise equilibrium and uses speed to move from one point to another. If there is no speed then the vehicle is a tethered aerostat or an untethered balloon"* [44].

Joner and Schneider [45] have categorized such partially buoyant concepts based on the *STOL* and *VTOL* category. STOL category was further divided into lifting body systems and auxiliary wing systems. Aereon, Dynaairship and Megalifter are few examples of such systems. These systems were probably designed to have higher *L/D* ratio as compared with the multi hull concept. They further divided the *VTOL* category into the lifting body systems and combined/ integrated systems. We are of the view that any aerial vehicle can be hybrid in terms of the propulsion system as well. The aerodynamic lift is basically used by the hybrid buoyant vehicles to meet the additional lift requirement at the takeoff and in the cruise-climb segment, including the steep turns. Moreover, in such configurations, weight of the airship and its payload are supported by the buoyant lift as well as by the aerodynamic lift. It is difficult to keep the gross takeoff weight as constant due to the continuous burning of the fuel during flight. Moreover, similar to a symmetric airfoil, conventional hull of the airship also provides aerodynamic lift, even at a small angle of attack. Therefore, defining the term hybrid in terms of the aerodynamics may be improper and defining it as hybrid buoyant aerial vehicle will avoid the fusion of hybrid propulsion technology with aerodynamic and hydrodynamic technologies. Hence the term "buoyant" is more suitable for aerial vehicles designed on eth Archimedes' Principle. Furthermore, this definition can be extended to cover all types of airships and aircrafts in a broader spectrum.

Wind Tunnel models of Goodyear airship [34] and that of the Akron airship [9] were the two major industrial wind tunnel testing performed for open jet conditions. Such a testing facility is sometimes referred as Eiffel type wind tunnel. It is usually consists of eight parts i.e. inlet, settling chamber, contraction cone, test section, diffuser, drive section and exhaust outlet part. In open jet testing, there is no influence of the wall as it is there for closed circuit tested section. In this way, the boundary layer of the walls also does not affect the results. Moreover, the scaled down models can be manufactured for larger scale factor as compared with those used inside the test section. In this way, the Reynolds number can be increased.

In recent years, Zhiyuan-1 a demonstration stratospheric airship and ESTOLAS (Extremely Short Take Off and Landing on Any Surface) aircraft model [46] were tested in open jet condition, shown in Figs. 2 and 3 respectively. ESTOLAS is a novel aircraft design which combines the best features of an airship, a plane, a helicopter and a hovercraft. Zhiyuan-1 is a demonstration, stratospheric airship by Shanghai Jiaotong University [18]. The total length of airship is 25m and that of its scaled down model is 1.8 m. Full scale model can occupy a volume of 750 m^3. The range of flight's Reynolds number is 1.8 to 9.3×10^6 and its model is tested at Reynolds number equal to 3.2×10^6 [18].

Fig. 2 Zhiyuan-1 airship model [18]

In subsonic wind tunnels with large test sections, scaled down models are attached to the balance with the help of model support system. Such a system mainly includes a main strut and an auxiliary strut, also known as pitch rod, Fig. 4. Such arrangement is different from that used in the case of open jet testing, shown in Figs. 2 and 3.

Fig. 3 ESTOLAS model for tube test, courtesy of A. Gamaleyev

The configuration shown in Fig. 4 is of a scaled down model of a HB aircraft to be tested in a closed-loop wind tunnel with a test section of dimensions 1.5m × 2.3m × 6m and maximum speed of 50 m/s [18]. In this tunnel, the balance section is placed on the floor to have some height for the external balance. Both the struts (main as well as auxiliary i.e. pitch rod for alpha mechanism) are attached with the balance. Whereas, the balance is usually housed in the strut in the case of internal balance attached with the model at its end. The configuration shown in Fig. 4 has fuselage structures made in machined aluminum. The model is supported by the strut at its central fuselage. It will then have an interface allowing pitch angle setting. The shape of the profile will be guided using ribs in machined aluminum, placed at a regular distance to ensure the exact contour. The wings and the canards are made in a composite, reinforced by structure in aluminum.

For *HBAVs*; the position of the gondola is subject to the requirement of the position of center of gravity to fulfill the desired static longitudinal stability criteria. For wall bounded testing, the position of the gondola may interact with the position of the strut. Hence, its contribution towards the overall aerodynamic and the stability response of the vehicle can be considered separately. Perhaps its position is also dependent on the stability results of clean configuration obtained from wind tunnel experimental. In case of the landing gears attached to the gondola, landing gears may not be modeled for the sub-scaled models, tested in subsonic wind tunnel of conventional test section size. One of the prospective reasons is that it is not possible to reproduce such small parts by using the same scale factor used for scaling hull, wings, horizontal and vertical tail.

Defining the scaling criteria for the manufacturing of the wind tunnel models for open jet testing is quite challenging. The term "scaling" is usually defined for a wind tunnel model to determine a scale down factor, without washing out the minute geometric details of the actual configuration. *HBAVs* need to be scaled as per the volume of hull and span of the wings and empennage. For such vehicles the thickness of the wing and empennage and overall weight of the model are the limiting factors for scaling the wind tunnel models. The scale down factor for *HBAV's* open jet testing models is more than those tested inside the test section of the wind tunnels.

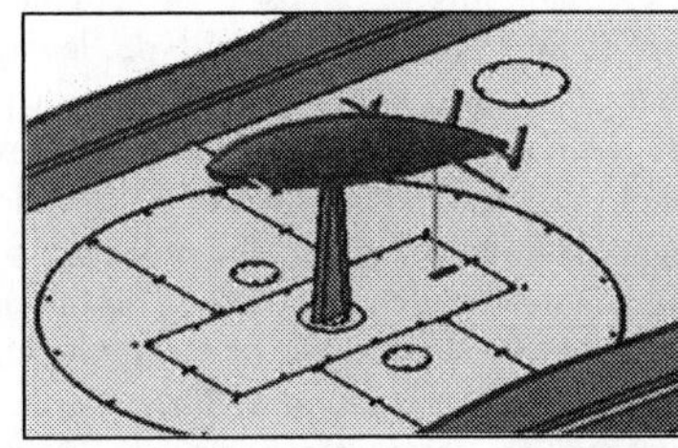

Fig. 4 Isometric view of HB aircraft installed in test section

II. Model Manufacturing

Manufacturing of the subscaled models should be ditto copy of full scale model. However, there is always requirement of additional screws for the mating parts. For example, long screws are required for the attachment of the wing and empennage with the main hull body. In such models, the empennage will act as a cantilever beam as one of its end is housed within the fuselage. This requires complete fluid structure analysis to first check the deflection of the aerodynamic surface under severe flow conditions during testing.

The far most important thing in wind tunnel model manufacturing is the selection of suitable scaling factor, which should be optimized as per the wind tunnel capabilities. Not only large, even minute details have to be exactly copied/shown in the scaled down wind tunnel model for the accuracy of results e.g. flushing of thin wings and voluminous fuselage body. For such models, not only different parts are required to be manufactured separately, some individual parts also have to be divided in sub parts. This has to be carried out sense fully and tactfully so that the integration of the parts of the model may be easily carried out without any problem and errors. For example, the wind tunnel model of *YEZ-2A* airship [47] is carried out using different combination of angles of these control surfaces . It can be observed from this testing and few previous testing of aerostats and airship that the model was rotated to 90 degree, such that the yaw axis becomes the pitch axis and vice versa. The basic reason behind this rotation is to avoid the interference of external strut. However, such a rotation will not be possible of *HBAVs* in which a wing is attached to the voluminous hull. Weight of the mode is an important consideration to cater the threshold values of the balance. In fact, using the full cross-section of the wind tunnel and keeping the weight up to a certain limit is quite a difficult job due to the voluminous fuselage. However, this issue becomes more critical for models to be tested in open jet testing due to comparatively higher scaling factor than those tested inside the test section.

For composites manufacturing, specially of the hull, the Vacuum Assisted Resin Infusion *(VARI)* or Light Resin Transfer Molding *(LRTM)* are the two most commonly methods used in the model manufacturing. In case of *VARI* it is an open molding process thus requiring only one contact surface but results in lesser structural strength and rigidity as compared to *LRTM* that is a closed molding process and requires upper and lower molds. For commercial and industrial applications, once the prototype is successfully qualified, filament wound composite structure can also be considered as it gives the maximum structural strength and rigidity to such structures of revolution. Regardless of the composite manufacturing technique, the mold making itself is the most critical and challenging task. Careful design and analysis is required to determine the optimum thickness of the composite structure and orientation of fiber angles. Typically for such structures, an intermittent sequence of glass fiber matte and unidirectional composite fabric is used for the manufacturing purposes. This also requires a detailed stress and strength analysis of composites structures using some finite element code for design and evaluation purposes. Finally the composite layup needs to be cured as per requirements of the used epoxy. It can either open to air curing or in controlled conditions depending on the size of mold and available curing facility.

Different approaches were available for integration of the control surfaces with the model. One of the prospective approaches is that the control surfaces be installed on the wings and tail with the help of different angle plates. A major problem is that the thickness of the control surfaces after scaling down was very small. Another problem was that these control surfaces had to be adjusted for different angle plates repeatedly and many times. The control surfaces and the plates are usually made of aluminum to reduce weight, but it is not possible to use steel screws because of repeated mounting and dismounting. This is because repeated use of steel screws in aluminum would wear out the threads in aluminum very soon. Thus the angle plates and control surfaces needs to be manufactured from steel. An alternative option is to use servos for controlling the deflection of control surfaces. Open jet testing serve this purpose efficiently and testing done for ESTOLAS model [46] is one of the example of it.

Surface finish of the wind tunnel models is very important for data quality. Lapping is one of the prospective option for metal parts. However, thee lapping process do not remove the burrs, and is ineffective in the round and the sharp edges. An alternative option is to make parts in composite, reinforced by structure in aluminum.

III. Static Testing

Talking of limitations related to open-jet wind tunnels, a major point of concern is the static pressure fluctuations in the test chamber. These static pressure fluctuations can not only affect the simulation quality, but can also limit the free stream velocity in the test section. It is well known that in open jet testing facilities, fluctuations in the static pressure is the limiting factor of the speed envelope and it also affects the flow quality. Large scale vortex structures are shed from the edges of the nozzle and jet length needs to be adjusted before conducting any experiment [48]. Moreover, similar to the standard wind tunnel testing, the open jet data is also subject to the correction and is affected by the model-nozzle and model collector interference [49]. These corrections depend upon the particular tunnel in which the model is tested, and are made necessary chiefly by the effects of sale, turbulence, and jet. Furthermore, add-on devices can perhaps help in decreasing the pressure at the exit plane of the nozzle [50].

Open jet testing has been traditionally known as open throat testing. The results of obtained by such testing are always quite in good agreement with the closed throat testing. An example of the same has been shown in Fig. 5 in which results of these two types of testing were compared for the Goodyear Zeppelin airship's hull model and were found to be within 8 percent. However, deviation in the results is observed at low Reynolds number cases and air stream turbulence is perhaps the basic reason for such deviation. Fig. 5 has been taken from [34] and used with permission of NASA. Furthermore, Increasing the turbulence appeared to increase the value of the minimum drag and maximum lift to drag ratio [51].

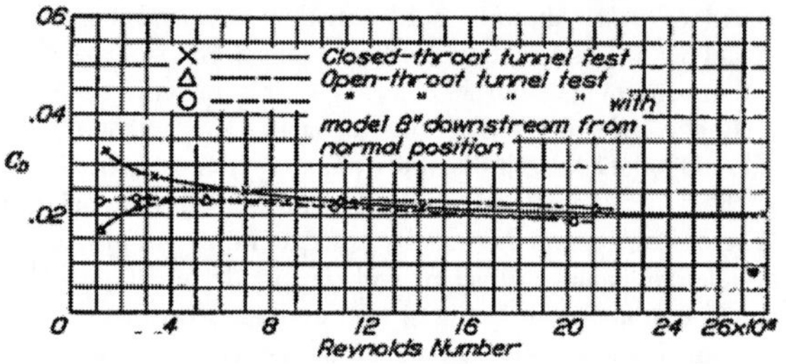

Fig. 5 Drag coefficient of Goodyear Zeppelin airship model. Fineness Ratio, 4.8; zero pitch; bare hull [34]

Open jet testing has been performed to study the effects of the surface roughness of the complete geometry of airship, Fig. 6 [18]. Surface roughness was introduced by employing the turbulence strips at three different locations. In this figure, hollow circles and solid squares represent the aerodynamic data with and without strip respectively. Whereas, the solid triangles symbolise the computational data obtained by using Fluent ®, a commercial Computational Fluid Dynamics analysis software.

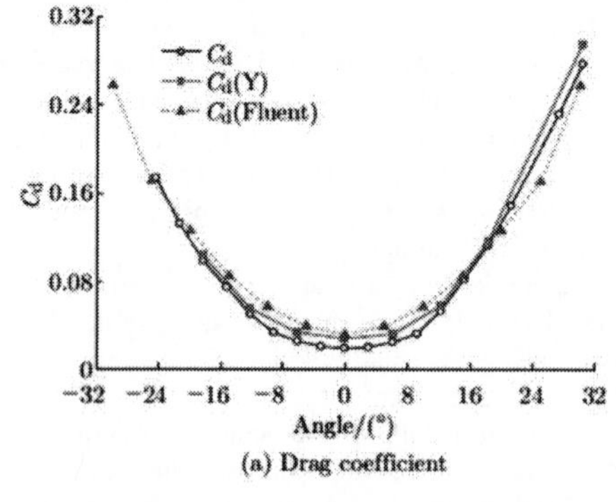

(a) Drag coefficient

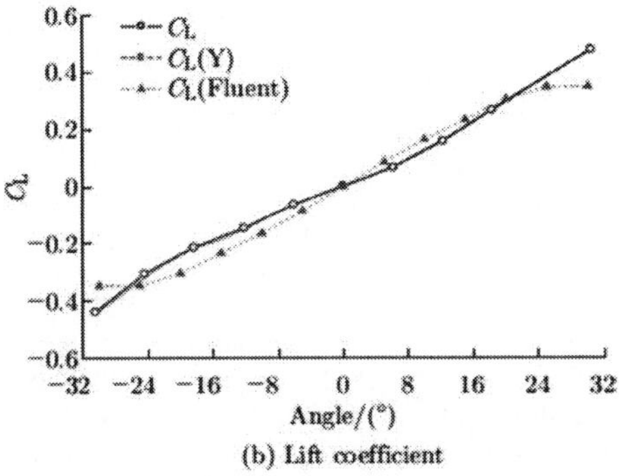

Fig. 6 Aerodynamic coefficients of Zhiyuan-1 airship model [18]

It can be observed that the turbulence strips had affected the drag till α equal to 15° and no obvious difference in results was observed for angles greater than this. The lift coefficient curves increased "*monotonously*" with increase in the pitch angle and no stall characteristics were captured for the defined range of pitch angle. Moreover the computational results were found to be in good agreement with those obtained from the wind tunnel testing.

Similar to airship models in which empennage are arranged at different angles on the hull, for hybrid buoyant aerial vehicles different empennage positions can be marked on the aft position of hull; thus providing a range of angle settings for moment and force measurement tests related to the study of the stability response along different axis. *ESTOLAS* model is one of the examples in which aerodynamic effects of the different settings of the flap deflection were experimentally studied, Fig. 7. Relative to the aircraft centerline, the stall angle is limited to 7° due to the use of high lift airfoil and incidence angle of 5°. The value of the lift coefficient and the corresponding stall angle were found to increase with increment in the deflection angle of the flaps. A shift in the drag polar at the positive deflection angle of the flaps was also observed due increase in the coefficient of drag.

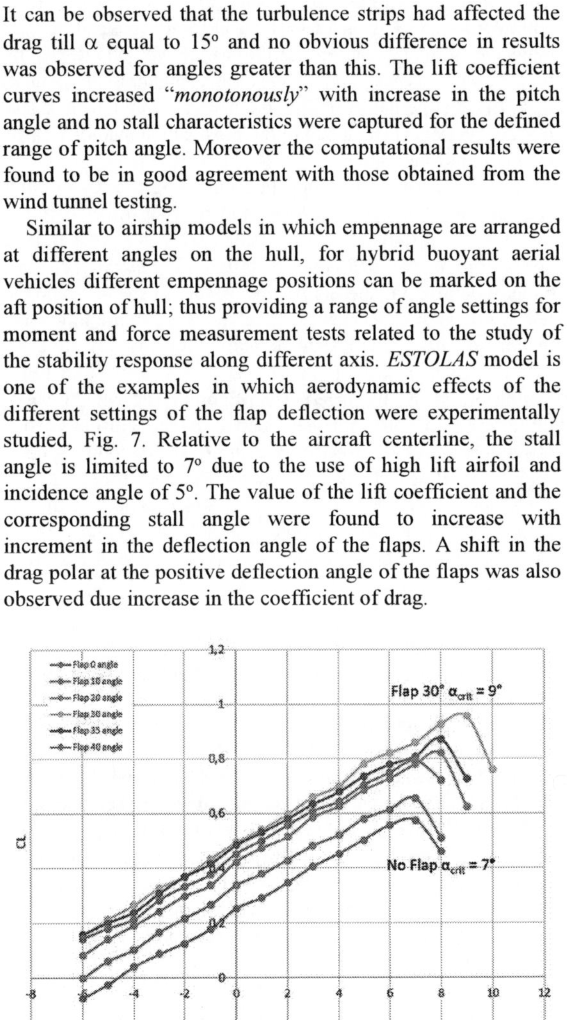

Fig. 7 $C_L$vs α plot of ESTOLAS Model with flap deflection [46]

IV. Discussion on Future Testing

One of the things common in the history of wind tunnel and airship is the time when design and development work was started. The first wind tunnel which was designed and developed was for the use of airship models [16]. However, unfortunately in past, more efforts were diverted towards aircraft technology and wind tunnels were design mainly to meet the requirements of the scaled down models of aircraft during world war–II [17]. The repurcation of this shortfall was badly felt in the last decade in which different buoyant and hybrid buoyant vehicles could not be tested inside the test section and open jet testing facilities were used for the said purpose [18]-[21]. For such vehicles, the drawback is that there might be the pressure loss due to the interaction of the flow with atmosphere [22]. If such models are tested inside the test section, then experiments can be performed under well controlled flow conditions [23]. However, the geometric scaling issues are quite difficult to handle, especially when a thin wing is attached to the voluminous fuselage.

Trim analysis (including hover and steady state), gust analysis (high frequency disturbances introduced by atmospheric turbulence), dynamic stability response (natural as well as forced), effect on aerodynamic parameters due to motion and inflation/deflation of the ballonets with shift in center of gravity and aeroelastic effects needs to be cater in wind tunnel testing. Especially, the aerodynamic results of flexible body of unconventional shaped *HBAVs* can be employed for the design & development of control system. It is due to the fact that the structural deformations (shear and torsional) will influence the aerodynamic forces. Another important problem of steady-state aeroelasticity is reversal, where the aeroelastic effects tend to decrease the lift of a lifting surface under certain circumstances. The typical example is an airfoil with a control surface attached to it. If the control surface is rotated in order to increase the lift, it is possible that the lift in the control surface itself will tend to twist the nose of the main airfoil down, causing a net decrease in the total lift. An essential area of research for experimental testing of *HBAVs* is also of *Dynamic Aeroelastic Instability* in which the flutter of the overall structures is there due to small disturbances induces more or less violent oscillations. Consider, for example, a thin wing attached to a voluminous fuselage of *HBAVs* and is tested in a wind tunnel. Usually, it is observed for the case of the wing that for low speeds of the flow, a perturbation of the wing sets an oscillatory motion, which is gradually damped. However, the same is not true at high subsonic speed. In this case, the wing is oscillatory unstable and is said to flutter. Hence, there is a need to conduct experiments to show that the oscillation is self-sustained, i.e., no external driving force is necessary, indicating that energy is being extracted from the flow. Closely related to the *Aeroelastic Stability Problems* for *HBAVs* are the response problems, in which the response of an aeroelastic system to an externally applied load is to he found. The external load may be caused by deformations of the elastic body or by disturbances in the fluid flow. The quantity of interest may be the displacement, the motion, or the stress state of the elastic body. In the case of the flutters, the response to a finite disturbance becomes indefinite. The major distinction between the response and stability problems is in the linearization used in the mathematical equations describing the physical problem. For stability analysis, the disturbances are usually regarded as small, thus the small deformation theory of elasticity and the linearization of the governing equations is justified. However, for response problems, it is necessary to consider finite deformations, since in this case we are interested not only in the modes of deformation but also in the absolute magnitude of the deformation and stress.

In past, wind tunnel testing on scaled models of airships has provided useful data on the aerodynamics and stability characteristics of hulls of different fineness ratio and its interaction with different empennage arrangement to check the control surface effectiveness. However, such tests are unique in nature as the Reynolds number (based on the cube root of the volume as reference area) is quite high. But, similar to any aircraft, if the Reynolds number is based on mean geometric cord, than it will be limited greatly due to the fact that the Reynolds numbers during the wind tunnel tests is small as compared with real flight one. For a fixed dimension of test section size of a subsonic wind-tunnel, Reynolds number is increased by increasing the flow velocity, which is limited by maximum achievable velocity by the fan of the wind tunnel. Like conventional aircraft, aerodynamic data obtained from wind tunnel for hybrid buoyant vehicles needs to be extrapolated to cater for the high Reynolds number effects. Moreover, it should be corrected from flight test data as well. Hence, all the necessary corrections may be catered far before using the wind tunnel data for engineering design purposes.

Force and measurement data of *HBAVs* at low speed is quite limited. Operational preparedness for its commercial application and its utilization to explore flow physics will remain incomplete till the time one has the complete matrix of data comparison as per the operational envelope of the wind tunnel. Moreover, as per general practice, low speed wind tunnel results may not be declared successful without discussing the margin of error on the measured values of aerodynamic forces and without justifying the reason for tunnel to tunnel variations in the aerodynamic and the stability data. In this regard, an effort can be done to test a standard *HBAVs* model at different testing facilities. It is perhaps not a new practice and LZ-120 airship's model was among the first, tested in two different wind tunnels [35]. Hence, the wind tunnel testing on similar model of *HBAVs* in different wind tunnels can birth to standard calibration models for wind tunnels.

During the flight test, a series of sensors are placed at the point where the *HBAVs* are firstly affected by the incoming air i.e. at the nose of the voluminous fuselage. The readings of these sensors are required to be correlated with the experimental pressure testing, which is perhaps quite limited for *HBAVs* in the open literature. But the same is not true of the case of airships as limited number of references are available [18], [22], [52]. Pressure distributions over the aerodynamic profile of the configuration are useful for the study of the aeroelastic behavior of the hull as well. It is perhaps an input for the central problem in steady-state aeroelasticity for which one has to estimate the effect of elastic deformation on the lift distribution over lifting surfaces such as wings of *HBAVs* at low speeds. Additionally, the estimation of the total drag and zero lift drag coefficient through wind tunnel testing of fuselage alone testing will be quite interesting, especially when such shapes can also generate some percentage of the overall aerodynamic lift. These data's can be utilized to estimate the contribution of the body towards the induced drag as well. In the case of the commercial aircraft, this type of drag is a matter of concern for environment and fuel saving. Worldwide, winglets are used in more than 5,000 individual airplanes and in 20 different types of aircrafts to reduce the induced drag [24]. By using blended winglets in jet aircrafts, about three billion gallons of jet fuel is jet fuel which in return reduced the CO_2 emission by more than 32.2 million tones [25]. Furthermore, to the author's best knowledge, the estimation of damping derivatives is perhaps missing and use of Damping Pitch Rate *(DPR)* apparatus [16] is one of the prospective methods for the estimation of the same.

Authors will not be wrong in mentioning that the airships are back in the form of *HBAVs* and it is the time to further explore the different testing techniques for unconventional configurations of *HBAVs*. In order to generate the real experimental data; the continuing trends in the provision of test facilities in experimental aerodynamics require innovative testing of *HBAVs* with data reliability for cost effective and safe design and development of such vehicles. Wind tunnel testing did so far on different configurations have shown the importance of the precise model manufacturing, which absolutely relies on precise machining of the components of scaled model of *HBAVs*. In some models, the empennage will act as a cantilever beam as one of it end is housed within the fuselage. This requires complete fluid structure analysis to first check the deflection under severe flow conditions which may the model with encounter during testing.

VI. Conclusion

In the light of future demand of economical hybrid airships for transportation of agricultural product and for tourism industry, it is anticipated that once the design of *HBAVs* get matured then, there will be a need to do the wind tunnel testing of its scaled down models at subsonic speed. Based on the initial literature survey, it is concluded that there still a gap in open jet testing of *HBAVs*. For example, the flow diagnostics testing to capture the vortical flow over aerodynamic contour of *HBAVs*. Nonetheless, the continued discovery of new knowledge in wind tunnel testing of *HBAVs* is critical for its future design and development.

References

[1] R. Pannel, J. R. and Jones, "Experiments on a Model of the German Rigid Airship L33,"NACA Report, 1917.

[2] A. H. Jones, R., Williams, D. H., and Bell, "Experiments on a Model of the Airship R.29," NACA Report, 1920.

[3] Jones, R., Williams, D. H., and Bell, A. H., "The distribution of pressure over the hull and fins of a model of the rigid airship R.101, and a determination of the hinge moments on the control surfaces.," NACA Report, 1920.

[4] V. Gomes and D. J. Speirits, "Airship response to turbulence: Results from a flight dynamics Simulation combined with a wind tunnel," AIAA Lighter-Than-Air Systems Technology Conference, 9 th, San Diego, CA, Technical Papers. Vol. 9. No. 11. 1991.

[5] Y. M. Huang and S. W. Changt, "Practical Design of an Airship," Journal of aircraft 32.6 (1995): 1294-1296.

[6] N. S. Tureaud, T.F., and Smith, "Wind Tunnel Characterization of a Scaled Class IV Aerostat," in *12th AIAA Ligher Than Air System Technology Conference*, 1997.

[7] P. Gill, S. Malik, and R. S. Pant, "Estimation of aerodynamic characteristics of un-symmetrically finned bodies of revolutions Estimation of Aerodynamic Coefficients Estimation of Unknowns Involved in Aerodynamic Coefficients," in *28th National Conference on Fluid Mechanics and Fluid Power*, 2001.

[8] A. Malik, S., Gupta S., Sharma, V., Purkait, P.K, and Singh, "Drag measurement through wind tunnel testing for different hull shapes used in aerostat application," Aerodynamic and Mechanical Decelerations and Inflatable Systems. Para, India, 1999.

[9] S. Sundaram, "Wind Tunnel Tests on 1:7 and 1:28 Scale Aerostat Models," PD EA 9905, Bangalore, India, 1999.

[10] S. P. Badesha, S., and Jones, "Aerodynamics of the TCOM 71 mm Aerostat," 1993.

[11] X. Wang, "Experimental Investigations on Aerodynamic Characteristics of the ZHIYUAN-1 Airship," Journal of aircraft (2010): 1463-1468.

[12] G. E. Dorrington, "Drag of Spheroid-Cone Shaped Airship," Journal of aircraft, (2006), 43(2), 363-371.

[13] T. Lutz, P. Funk, A. Jakobi, and S. Wagner, "Summary of Aerodynamic Studies on the Lotte Airship" In 4th International Airship Convention and Exhibition, pp. 1-12. July, 2002

[14] P. Funk, T. Lutz, and S. Wagner, "Experimental investigations on hull-fin interferences of the LOTTE airship," *Aerospace Science and Technology* 7.8 (2003): 603-610.

[15] C. Jordi, S. Michel, and E. Fink, "Fish-like propulsion of an airship with planar membrane dielectric elastomer actuators.," *Bioinspiration & biomimetics* 5, no. 2 (2010): 026007

[16] A. Ribeiro, V. R. Mantha, A. I. Rouboa, D. a. Marinho, and A. J. Silva, "Study of external air flow for an AURORA," *Aircraft Engineering and Aerospace Technology* 83, no. 5 (2011): 255-265.

[17] S. C. Chan and J. D. Hunt, "Wind Tunnel Study of a Large Aerostat," *AIAA Paper* 7068 (2011): 20-22.

[18] P. Liu, G. Fu, L. Zhu, and X. Wang, "Aerodynamic characteristics of airship Zhiyuan-1," *Journal of Shanghai Jiaotong University (Science)* 18 (2013): 679-687.

[19] B. T. Buerge and I. Introduction, "Influence of Surface Roughness on the Drag of Very Large," AIAA lighter-than-air systems technology (LTA) conference 2013, no. March, pp. 1–6.

[20] S. Chan, K. Shervington, and J. Hunt, "Wind Tunnel Study of a Large Aerostat, CFD Validation," *AIAA Lighter-Than-Air Systems Technology (LTA) Conference*., pp. 1–8, March 2013.

[21] P. Liu, J. Sun, and Z. Tang, "Wind Tunnel Tests of the Counter-Rotating Propeller of Stratospheric Airship," In *APCOM & ISCM Conference*, 2013.

[22] P. Liu, Y. Fu, L. Wang, and Q. Song, "Study on Pressure Coefficient Distribution of the Airship Zhiyuan-," Journal of Harbin Institute of Technology 3 (2013): 015..

[23] K. E. Urbahs, A., D. Titovs, A.Barovs, A. Aleksandrovs, S. Lučkinskis, "Experimental Research of aerodynamic characteristics of the ESTOLAS hybrid aircraft," in *AVIA INVEST International Conference*, 2014, no. October.

[24] A. D. Andan, W. Asrar, and A. a. Omar, "Investigation of Aerodynamic Parameters of a Hybrid Airship," *Journal of Aircraft* 49, no. 2 (2012): 658-662.

[25] Y. Liu, P. Zeng, and L. P. Lei, "Experimental Study on the Stability Properties of Different Design of Tandem Wing Airship Models," *Appl. Mech. Mater.*, vol. 457–458, pp. 1611–1614, Oct. 2013.

[26] A. Suñol, D. Vucinic, and S. Vanlanduit, "Experimental and Numerical Study of the Effect of Lateral Wind on the Feeder Airship," proceedings of International Conference on Fluid Dynamics and Mechanics (ICFDM-2013). 2013.

[27] R. Rist and B. Martin, "Introduction of the Dynalifter ™ Aircraft." 2012. (Online). Available: tmacog.org/OCF_09/Presentations /Dynalifter.pdf. Accessed November 30, 2015

[28] Y. Chen, P. Liu, Z. Tang, and H. Guo, "Wind tunnel tests of stratospheric airship counter rotating propellers," *Theor. Appl. Mech. Lett.*, Feb. 2015.

[29] A. U. Haque, N. Mohamad, A. Hafiz, S. M. Kashif, A. A. Omar, E. Sulaeman, and J. S. M. Ali, "Design and Fabrication of a Winged Hybrid Airship Model for IIUM-LSWT," In *Advanced Materials Research, vol. 1115, pp. 513-516. 2015.*

[30] *F. Dynamics* and P. Worklng, *Quality Assessment for Wind Tunnel Testing*, July. 1994.

[31] J. Green and J. Quest, "A short history of the European Transonic Wind Tunnel ETW," *Progress in Aerospace Sciences* 47, no. 5 (2011): 319-368.

[32] Grant E Carichner and Leland M Nicolai, *Fundamentals of Aircraft and Airship Design: Volume II, Airship Design*. American Institute of Aeronautics and Astronautics,AIAA, 2013.

[33] H Multhopp "Aerodynamics of the Fuselage." NACA TM 1046, 1942.

[34] R. W. Abbott, I. H., & Noyes, "Airship model tests in the variable density wind tunnel," NACA Report, 1931.

[35] C. Stephenson,. Zeppelins: German Airships, 1900-40 (Vol. 101). Osprey Publishing, 2004.

[36] A. Gamaleyev, "ESTOLAS Project Description," 2012. (Online). Available: http://cordis.europa.eu/result/rcn/58762_en.html. Accessed December 1, 2015.

[37] R. Rist,"World's First Practical Airship," 2012. (Online). Available: tmacog.org/OCF_09/Presentations/Dynalifter.pdf. Accessed November 30, 2015.

[38] Hybrid Air Vehicles, Press Release, "Hybrid Air Vehicles receives a multi-million pound government grant, to ensure Airlander returns to flight, 2015, URL: http://www.blimpinfo.com/airships/ hybrid-air-vehicles-receives-a-multi-million-pound-government-grant-to-ensure-airlander-returns-to-flight/. Accessed December 11, 2015.

[39] A. U. Haque, W. Asrar, A. A. Omar, E. Sulaeman, and J. S. M. Ali, "Hydrostatic and Hydrodynamic Characteristics of Swimming Animals-An Inspiration for Hybrid Buoyant Aircraft." 4th ICASE, 3rd - 4th October, 2015, Langkawi, Malaysia

[40] B. Gumse, "Low-Speed Wind-Tunnel Tests of a Full-Scale M2-F2 Lifting Body Model," NACA Report, 1967.

[41] G. Ash, Lawrence, (Flight Test and Wind Tunnel Performance Characteristics of the X-24A Lifting Body, FTC-TD-71-8, June 1972

[42] E. and C. I. C. Jackson, "'Preliminary Subsonic Aerodynamic Model for Simulation Studies of the HL-20 Lifting body," National Aeronautics and Space Administration, Office of Management, Scientific and Technical Information Program, 1992 , 1992.

[43] B. T. Buerge, "Scalability of Heaviness Fraction for Large Airships," AIAA Lighter-Than-Air Systems Technology (LTA) Conference, Proceedings-20th Conference, 2013-1317, 2013..

[44] G. E. Carichner and L. M. Nicolai, "Hybrids ... the Airship Messiah ?," AIAA Lighter-Than-Air Systems Technology (LTA) Conference, Proceedings-20th Conference, 2013-1317, 2013.

[45] B.,Jones, D., Grant, H.Rosenstein, J. Schneider. Feasibility Study of Modern Airships, Final Report, Phase I, Vol. I. NASA CR-137691, 1975

[46] K.E. Urbahs, A., D. Titovs, A.Barovs, A. Aleksandrovs, S. Lučkinskis, Experimental Research of aerodynamic characteristics of the ESTOLAS hybrid aircraft, in: AVIA INVEST Int. Conf., Riga, Latvia, 2014.

[47] S. B. V., Gomes, "An investigation into the flight dynamics of airships with application to the YEZ-2A." (1990).

[48] M. Rennie, "Effect of jet length on pressure fluctuations in 3/4 open-jet wind tunnels. In Motor Industry Research Association, Vehicle Aerodynamics Symposium, October 2000

[49] S.,Oerlemans, L., Broersma and P.Sijtsma (2007). Quantification of airframe noise using microphone arrays in open and closed wind tunnels. International Journal of Aeroacoustics, 6(4), 309-333..

[50] C., Kramer, H. J. Gerhardt and L. J. Janssen, "Flow studies of an open jet wind tunnel and comparison with closed and slotted walls", Journal of Wind Engineering and Industrial Aerodynamics, 22(2), 115-127, 1986.

[51] M., Knight, T. A. Harris (1930). Experimental determination of jet boundary corrections for airfoil tests in four open wind tunnel jets of different shapes. US Government Printing Office.

[52] R.Hills R. A Review of measurements on AGARD calibration models. NACA Report, 1961.

Simulation Study on Vehicle Drag Reduction by Surface Dimples

S. F. Wong, S. S. Dol

***Abstract*—**Automotive designers have been trying to use dimples to reduce drag in vehicles. In this work, a car model has been applied with dimple surface with a parameter called dimple ratio *DR*, the ratio between the depths of the half dimple over the print diameter of the dimple, has been introduced and numerically simulated via *k-ε* turbulence model to study the aerodynamics performance with the increasing depth of the dimples The Ahmed body car model with 25 degree slant angle is simulated with the *DR* of 0.05, 0.2, 0.3 0.4 and 0.5 at Reynolds number of 176387 based on the frontal area of the car model. The geometry of dimple changes the kinematics and dynamics of flow. Complex interaction between the turbulent fluctuating flow and the mean flow escalates the turbulence quantities. The maximum level of turbulent kinetic energy occurs at *DR* = 0.4. It can be concluded that the dimples have generated extra turbulence energy at the surface and as a result, the application of dimples manages to reduce the drag coefficient of the car model compared to the model with smooth surface.

***Keywords*—**Aerodynamics, Boundary Layer, Dimple, Drag, Kinetic Energy, Turbulence.

I. Introduction

A. Problem Statement

As of today, the study on the effect of dimples focusses on bluff body application (i.e. golf ball) than streamlined body application (i.e. car surface). Several previous studies (experimental and numerical) however observed that dimples on an airfoil create extra turbulence to delay the boundary layer separation. For example, [1] has suggested a smart dimple matrix over the airfoil which will sense boundary layer separation and arrange dimple in the least drag and high lift configuration which is shown in Fig. 1. Moreover, composite shaped and outward dimples have the least drag when compared to the round and inward dimples. It was found that the dimples will reduce the velocity in the boundary layer and also reduce the surface vorticity. Besides that, dimples will also make the momentum layer inside the boundary become thinner leading to drag reduction [2], [3].

German researchers proved an effect which could revolutionize the design of ships, vehicles and aircrafts. They came to an unexpected result: the train with dimples has 16% less frictional resistance, depending on the speed. The finding seems contradicting the theorem that says surface roughness (dimple) increases the surface area thus friction. Moreover, the drag reduction was overcome by the newly appearing pressure force contribution due to surface unevenness. The pressure validation involved a Scanivalve unit and a Validyne differential pressure gauge. Fig. 2 shows a typical example for each test of Reynolds number *Re* ¼ $U_b d$=*m* (bulk velocity U_b, channel half-width *d*) the pressure gradient d*p* =d*x* was deduced by a linear fit of the measured pressure data [4]. Therefore, adding a dimple on a streamlines body might help to delay the flow separation and reduce the size of the wake but it might increase the friction drag as a trade-off. The present work aims to provide the answer with physics clarifications.

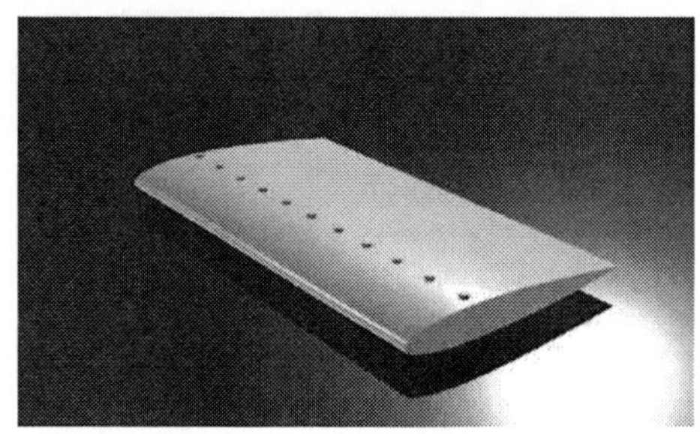

Fig. 1 Multi dimpled configuration

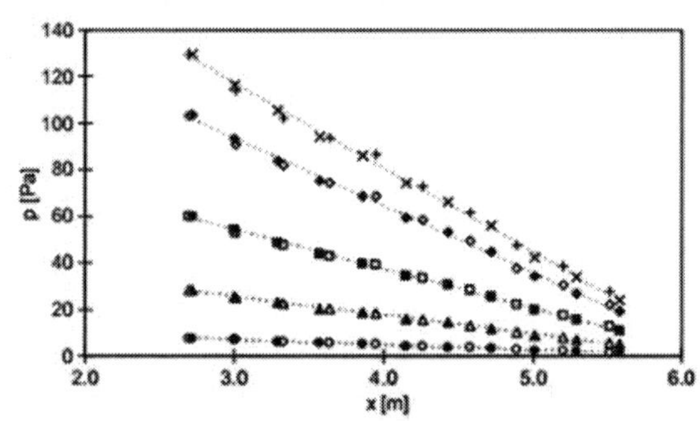

Fig. 2 Pressure distribution measured along flat and dimpled side of the channel for various Reynolds numbers

S. F. Wong and S. S. Dol are with the Department of Petroleum Engineering, Curtin University, Miri, Sarawak, Malaysia (e-mail: wong.siew.fan@postgrad.curtin.edu.my, sharulsham@curtin.edu.my).

International Science Index, Mechanical and Mechatronics Engineering Vol:10, No:3, 2016 waset.org/Publication/10003966

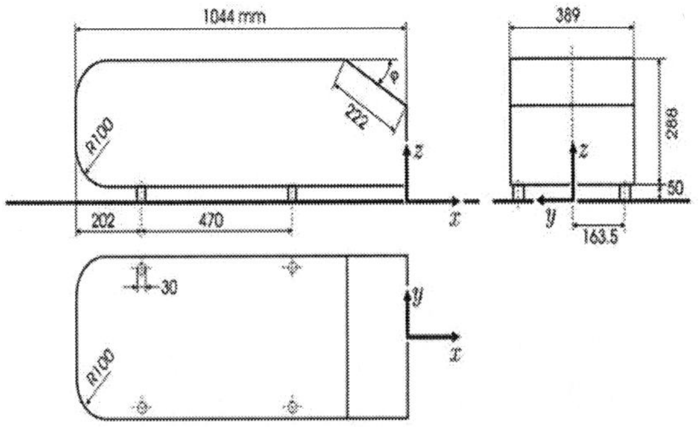

Fig. 3 Dimension details of the Ahmed body car model

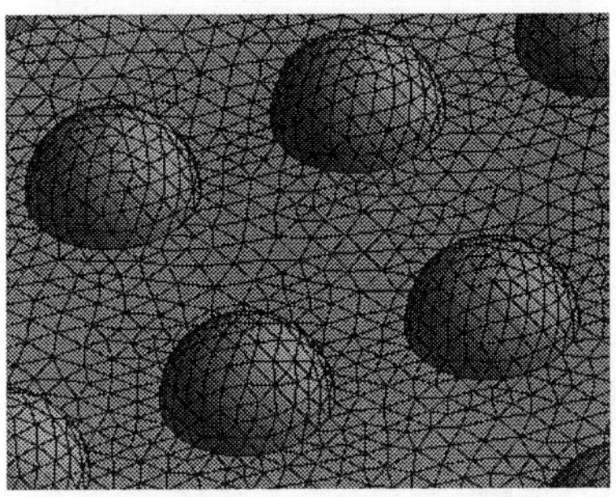

Fig. 4 Tetrahedral mesh around dimples

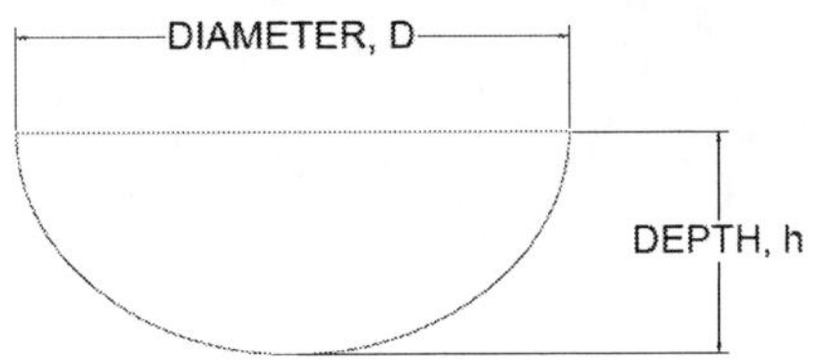

Fig. 5 Dimple ratio

TABLE I
QUALITY OF MESH FOR ALL MODELS

Model	Nodes	Elements	Skewness	Orthogonal Quality
Smooth surface	142564	782743	0.222	0.862
DR = 0.05	319366	1798765	0.221	0.863
DR = 0.2	282862	1578882	0.221	0.863
DR = 0.3	283411	1580105	0.221	0.863
DR = 0.4	284911	1586428	0.221	0.863
DR = 0.5	280090	1560387	0.221	0.863

TABLE II
BOUNDARY CONDITION

Boundary	Conditions
Ahmed Body	Stationary Wall
Sides	Symmetry
Top	Symmetry
Center	Symmetry
Road	Stationary Wall
Inlet	Velocity Inlet
Outlet	Pressure Outlet

II. APPROACH AND METHODS

In this work, Ahmed body car model with 25-degree slant angle will be used. Ahmed body car model is a simplify car model for accurate flow simulation but still retain with the significant features of a car which are curved fore body, straight center section and the angled rear end. As shown in Fig. 3, it is a typical bluff body which commonly using for the simulation to study the flow past a car [5].

For the computational fluid dynamics (CFD) simulation, the tetrahedral meshing is applied (Fig. 4) and k-ε model in ANSYS Fluent has been chosen for turbulence modeling. *k-ε* model consists of two transport equations to represent the turbulent properties of the flow. The first transport variable is the kinetic energy term, *k*. It is the variable to determine the turbulent energy production.

$$\partial/\partial t\,(\rho k) + \partial/\partial x_i\,(\rho k u_j) = \partial/\partial x_j\,[(\mu + \mu_t/\sigma_k).\partial/\partial x_j] + G_k + G_b - \rho\varepsilon - Y_M + S_k \quad (1)$$

The second transport variable is the turbulent dissipation term, *ε*. It is the variable to determine the scales of the turbulence.

$$\partial/\partial t\,(\rho\varepsilon) + \partial/\partial x_i\,(\rho\varepsilon u_j) = \partial/\partial x_j\,[(\mu + \mu_t/\sigma_\varepsilon)\,\partial_\varepsilon/\partial x_j] + \rho C_1 S_\varepsilon - \rho C_2[\varepsilon^2/(k + \sqrt{v\varepsilon})] + C_{1\varepsilon}.\varepsilon/k\,.C_{3\varepsilon}.G_b + S_\varepsilon \quad (2)$$

G_k is the generation of turbulence kinetic energy due to the mean velocity gradients. G_b is the generation of turbulence kinetic energy due to buoyancy. Y_M is the contribution of the fluctuating dilatation in compressible turbulence to the overall dissipation rate. $C_{1\varepsilon}$ *and* C_2, and σ_k *and* σ_ε are constants and turbulent Prandtl numbers for k and ε respectively. S_k *and* S_ε are user defined source terms. Further info can be referred in [6].

A parameter called dimple ratio (DR) has been introduced. Basically, dimple ratio is the ratio between the depths of the half dimple over the print diameter of the dimple. In this work, car model with be simulated with the DR of 0.05, 0.2, 0.3 0.4 and 0.5 at Reynolds number of 176387 based on the frontal area of the car model.

$$\text{Dimpleratio,DR} = \text{depth,}h/\text{diameter,}D \quad (3)$$

The meshing is done by using ANSYS meshing software. Tetrahedral meshing is used for the analysis. Table I shows the setting that has been applied to all models. It shows that the skewness and the orthogonal quality of the meshing are within the acceptable region of simulation accuracy.

In this work, the air velocity was set as 40m/s at the inlet. The other boundary conditions are presented in Table II.

The following equation has been used to calculate the Reynolds number of the flow in order to determine the flow past the model either is laminar or turbulent.

$$Re = \frac{\rho U A}{\mu} = \frac{1.2 \times 40 \times 0.065741}{1.789 \times 10^{-5}} = 176387$$

where: ρ is the density of air; U is the flow velocity; A is the frontal area of the car model; μ is the dynamic viscosity of air.

The value obtained for the Reynolds number corresponds to a turbulent air flow.

III. Result and Discussion

From Figs. 6-9, two main changes can be observed. First, pressure is reduced at the end of the roof which is the edge for all the models that have dimples. When compared with the model without dimple (Fig. 6), dimpled models have lower pressure at the edge. This is due to the accelerating velocity at the vicinity (e.g. Fig. 9 (*e*)).

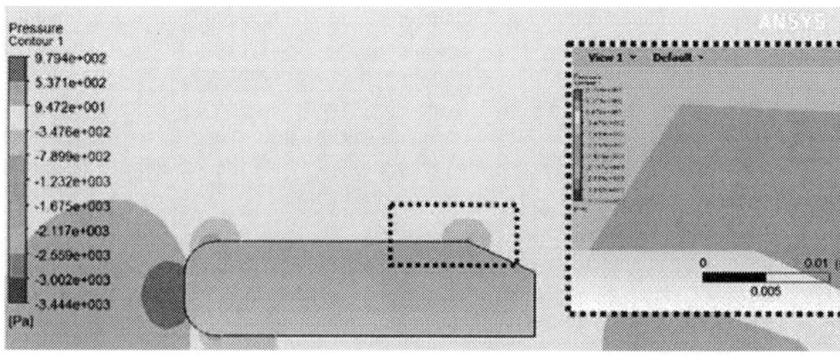

Fig. 6 Pressure contour for model with smooth surface

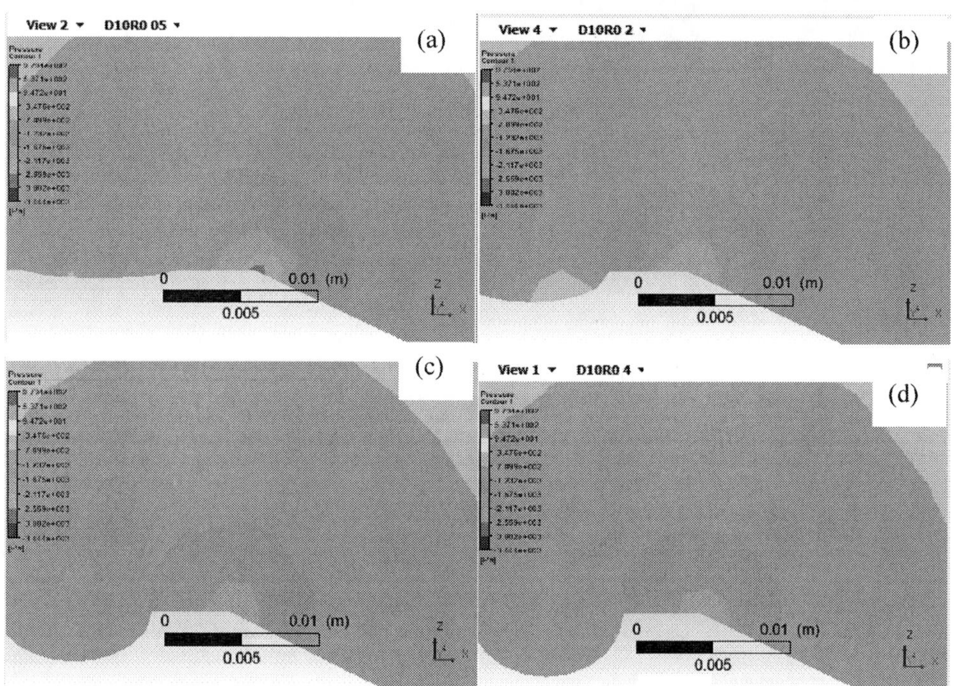

(e)

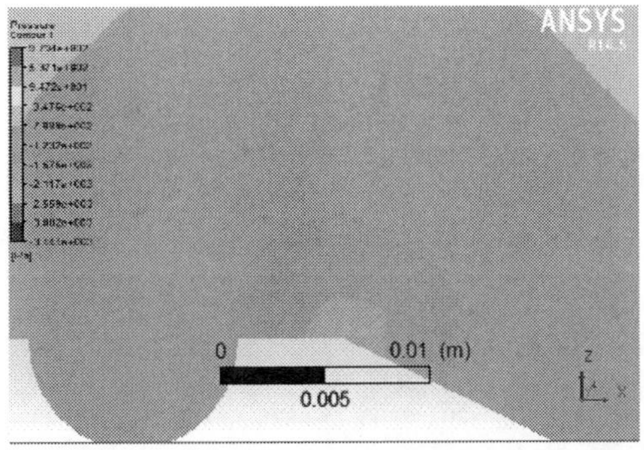

Fig. 7 Pressure contour (a) DR = 0.05; (b) DR = 0.2; (c) DR = 0.3; (d) DR = 0.4; (e) DR = 0.5

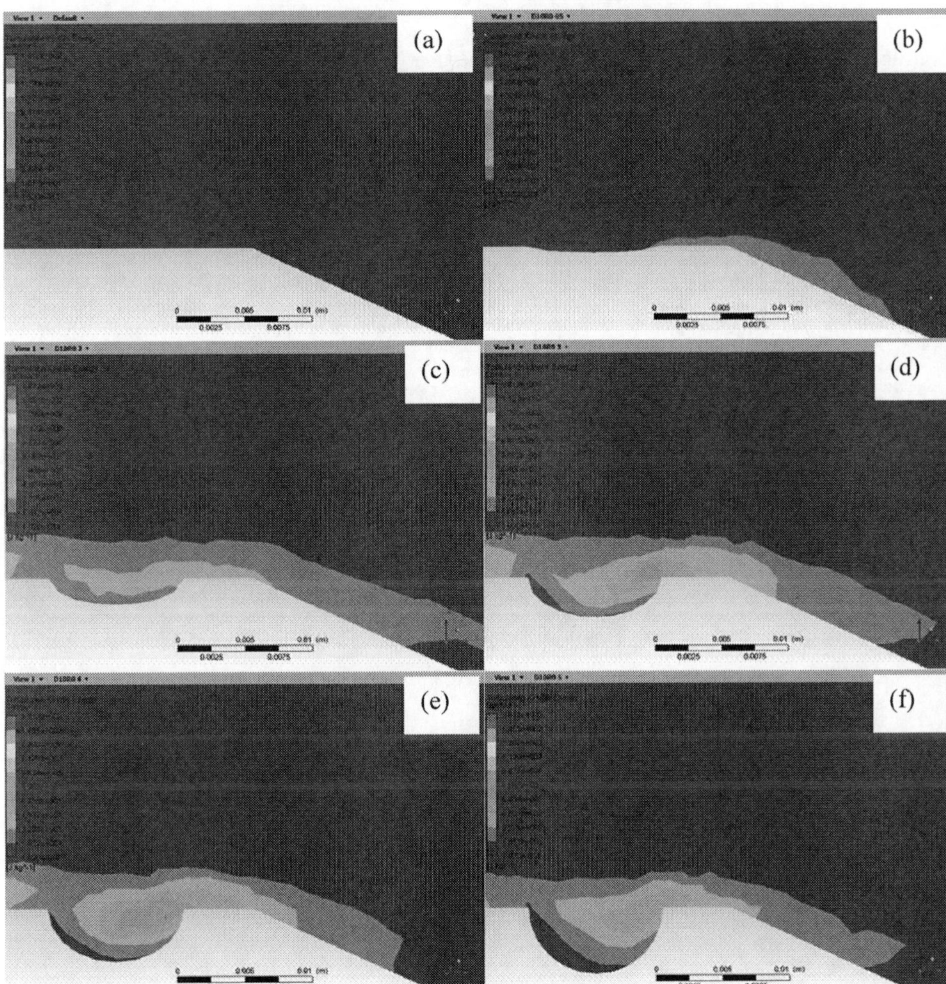

Fig. 8 Turbulence kinetic energy contour (a) DR = 0; (b) DR = 0.05; (c) DR = 0.2; (d) DR = 0.3; (e) DR = 0.4; (f) DR = 0.5

When compared with the model without dimple (Fig. 8 (a)), turbulent kinetic energy is generated within the dimple and at the vicinity of the dimple edge (Figs. 8 (b)-(f)). These results suggest that the boundary layer flows manage to go further before the flow separation takes place.

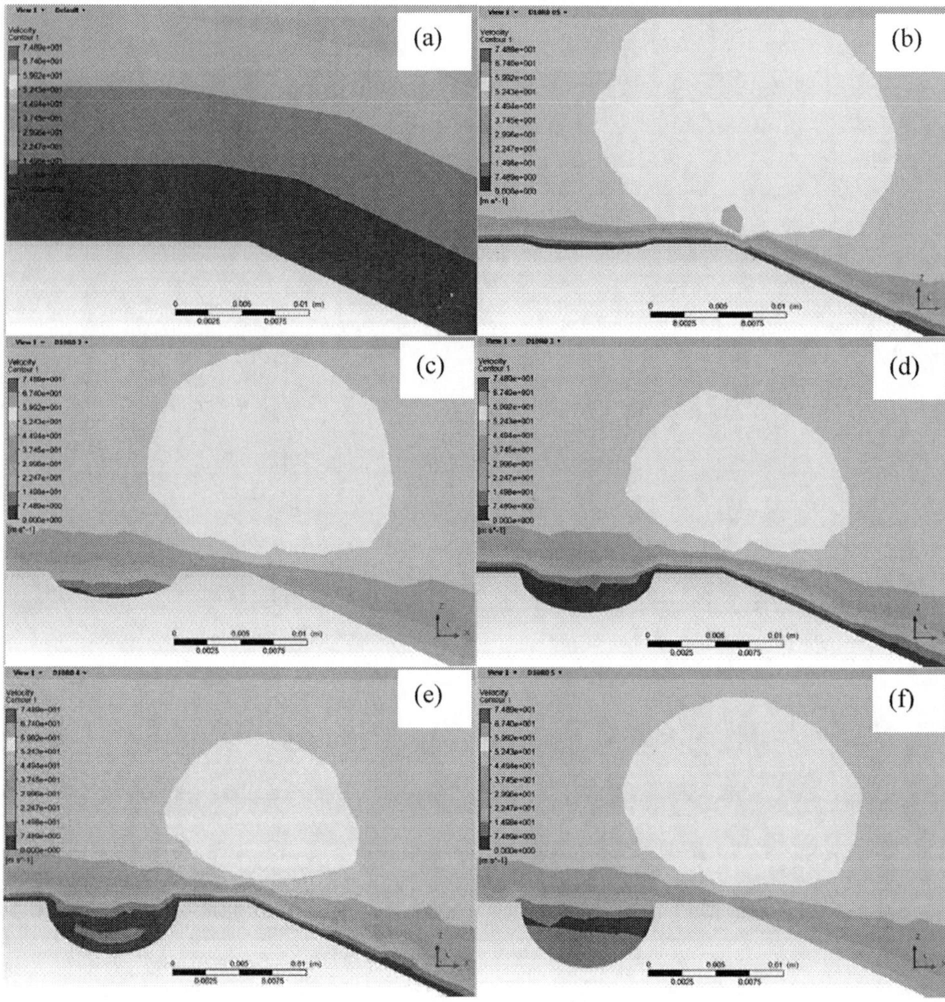

Fig. 9 Velocity contour (a) DR = 0; (b) DR = 0.05; (c) DR = 0.2; (d) DR = 0.3; (e) DR = 0.4; (f) DR = 0.5

The geometry of dimple changes the kinematics and dynamics of flow. Complex interaction between the turbulent fluctuating flow and the mean flow escalates the turbulence quantities and the boundary layer gains momentum from this turbulent kinetic production. The maximum level of turbulent kinetic energy occurs at DR = 0.4 as shown in Fig. 8 (e). From these changes, it can be concluded that the dimples have generated extra turbulence energy at the surface. With the extra energy, the flow manages to go further before the flow separation take place.

TABLE III
COMPARISON OF DRAG COEFFICIENT

Model	Drag Coefficient	Percentage Difference (%)
Smooth surface	0.261	-
DR = 0.05	0.257	1.32
DR = 0.2	0.259	0.55
DR = 0.3	0.260	0.07
DR = 0.4	0.255	1.95
DR = 0.5	0.258	0.84

One of the aims of this work is to determine how much changes the dimples will bring to the coefficient of drag. All the dimple ratio models give smaller coefficient of drag, C_D

compared to the model without dimple (Table III). The coefficient of drag, C_D is reduced by 1.95% for the model with DR = 0.4.

IV. Conclusion

Application of dimple on the car surface has proven to be effective. The dimples manage to reduce the drag coefficient. The results are encouraging since the simulation is only based on one dimple. Different parameters like dimples position, number of dimples, dimples orientation will be tested in order to fully understand the performance of the dimple application on vehicle aerodynamics.

References

[1] D. Srivastav, Flow control over airfoils using different shaped dimples, *International Conference on Fluid Dynamics and Thermodynamics Technologies*, (2012) 92–97, Singapore: IACSIT Press.

[2] L.-m Tian, L.-q Ren, Q.-p Liu, Z.-w Han, X, Jiang, The mechanism of drag reduction around bodies of revolution using bionic non-smooth surfaces, *Journal of Bionic Engineering*, (2007) 109-116.

[3] Dol, S.S., Azimov, U. & Martinuzzi, R.J.M. (2012). Statistical description in the turbulent near wake of rotating circular cylinder. World Academy of Science, Engineering and Technology 72, 505 – 509.

[4] L. Hermann, B. Michael, K. Cagatay, Drag Reduction by dimples? A Complementary Experimental/Numerical Investigation, *International Journal of Heat and Fluid Flow*, 29(3), (2008) 783-791.

[5] S.R. Ahmed, G. Ramm, G. Faltin, Some salient features of the times-averaged ground vehicle wake, *SAE Society of Automotive Eng., Inc*, 1(840300) (1984) 1–31.

[6] C.K. Chear, S.S. Dol, Vehicle Aerodynamics: Drag Reduction by Surface Dimples? *International Journal of Mechanical, Aerospace, Industrial and Mechatronics Engineering* 9(1) (2015), 202-205.

Aerodynamic Models for the Analysis of Vertical Axis Wind Turbines (VAWTs)

T. Brahimi, F. Saeed, I. Paraschivoiu

***Abstract*—**This paper details the progress made in the development of the different state-of-the-art aerodynamic tools for the analysis of vertical axis wind turbines including the flow simulation around the blade, viscous flow, stochastic wind, and dynamic stall effects. The paper highlights the capabilities of the developed wind turbine aerodynamic codes over the last thirty years which are currently being used in North America and Europe by Sandia Laboratories, FloWind, IMST Marseilles, and Hydro-Quebec among others. The aerodynamic codes developed at Ecole Polytechnique de Montreal, Canada, represent valuable tools for simulating the flow around wind turbines including secondary effects. Comparison of theoretical results with experimental data have shown good agreement. The strength of the aerodynamic codes based on Double-Multiple Stream tube model (DMS) lies in its simplicity, accuracy, and ability to analyze secondary effects that interfere with wind turbine aerodynamic calculations.

***Keywords*—**Aerodynamics, wind turbines, VAWT, CARDAAV, Darrieus, dynamic stall.

I. Introduction

WHILE human have been harnessing the energy of the wind for thousands of years the interest in wind power was reborn during the energy crises of the 1970s period in which the economy of the major industrial countries of the world, particularly the United States was affected by the 1973 oil crisis and the 1979 energy crisis. Since then many countries turned to renewable energy in particular wind energy where the power collected from wind machines is fed into the existing electricity grid, combined with electricity from other power plants, and finally distributed to utility customers. The J.-A. Bombardier Aeronautical Chair at Ecole Polytechnique has developed computer codes to predict aerodynamic loads and performance of a Darrieus rotor including viscous, stochastic wind and dynamic-stall effects [1]. A fundamental study of the dynamic-stall phenomenon has also been performed using the unsteady Navier-Stokes equations in a stream function-vorticity formulation [2], [3]. For fast and accurate predictions of vertical-axis wind turbines performances can be obtained using momentum models based on the conservation of momentum principle in a quasi-steady flow. The present paper reviews the main aerodynamic models that have been developed to predict the aerodynamic loads and performance of wind turbines.

T. B. is Assistant Professor with the Electrical and Computer Engineering Department, Effat University, KSA (e-mail: tbrahimi@effatuniversity.edu.sa).

F. S. is Associate Professor with the Aerospace Engineering Department, King Fahd University of Petroleum & Minerals, KSA (e-mail: farooqs@kfupm.edu.sa).

I. P. is Professor with the Mechanical Engineering Department, Montreal, Canada (e-mail: ion.paraschivoiu@polymtl.ca).

II. Development of Wind Energy in Canada

Early developments of wind energy in Canada were located primarily in Ontario, Quebec, and Alberta. Throughout the late 1990s and early years of the 21st Century every Canadian province has pursued wind power to supplement their provincial energy grids. Indeed, in Canada the energy is a provincial responsibility not a federal one. The federal government introduced a series of greenhouse gas (GHG) emission regulations which would support shifting to clean energy sources like solar, wind and geothermal, but industry is putting more efforts on provincial policies and programs that will help in improving the development of wind energy. In Quebec alone, more than 100,000 MW of wind potential exist in sites within 25 km of Hydro Quebec's existing transmission lines [4]. The number of wind turbines in Quebec exceeds 1600 turbines, with a total installed capacity of 2883 MW, and has one of the country's largest wind farm, the 350 MW at Rivière-du-Moulin. Quebec ranks second among Canadian provinces in terms of wind power installed powering 1,000,000 Quebec homes. The Global Wind Energy Council [5] reported that each year the average of new wind farm construction is about 1.4 GW, this trends is expected to be maintained for the next couple of years. CanWEA, the Canadian Wind Energy Association recently published the new vision for wind energy in Canada for 2025 discussing and showing how wind energy in Canada will create new jobs and reduce the impact on the environment [4]. Canada's electricity system can integrate large amounts of wind energy and by 2025 Canadians can get 20 percent of their electricity from wind that would require installing 22,000 wind turbines in 450 locations. In 2014, as of September 2015 Canada had a total wind installed capacity of nearly 10,425 MW, Fig. 1 [6]. The quality of Canada's wind resource is as good as or better than any of the world's leading wind energy nations such as Germany and the United States. For the coming four years, Canada will see an average of 1,500 MW of new projects creating 68,000 additional person-years of employment and attracting $15 billion in new investment, in addition, by 2025, Canada would reach a capacity of 55,000 MW, meeting 20% of the country's energy needs [4].

While the cost to build wind energy continues to decline, the improvements made in wind turbine through advanced technology had significant effects on the efficiency of the machines and become an important commercial option for large scale power production. Current turbine technology has enabled wind energy to enter the electric power mainstream and become a viable power source in today's energy market [4], [7], [8]. Over the past few years' novel designs are taking wind power to the next level where the efficiency is being gained through new models and new advanced technology. These improvements are made in many parts of the machine including gearbox, drive units, power electronics, better control algorithms and data acquisition, and more importantly in the aerodynamic of the rotors that improved blade technology using lighter, stronger, and aerodynamically performing and efficient rotor blades. The new generation of advanced wind turbines include blades made from soft, flexible materials that change shapes in response to wind speed or aerodynamic loads. The technology is part of a trend making wind power almost as cheap as fossil fuels. For example, in 1991, the cost of a kilowatt hour from wind power cost 15 cents while now it has dropped to 6.5 cents per kilowatt hour [7].

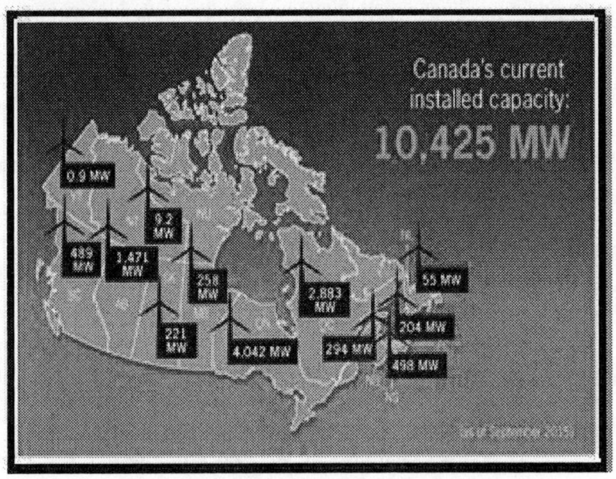

Fig. 1 Canada's current installed capacity [6]

III. Aerodynamic Models for VAWT

Modern wind turbines can basically be classified as either of vertical-axis (VAWT) design, such as the Darrieus model, or as horizontal-axis (HAWT) variety like the traditional farm wind pump, Fig. 2. In both cases, the turbine's rotating blades extract kinetic energy from the wind to generate electricity or pump water. Although the HAWT remains the main focus of all wind energy research and development, the vertical-axis wind turbine model offers an advantageous alternative due to its mechanical and structurally simplicity of harnessing the wind energy [1], [9]. The J.-A. Bombardier Aeronautical Chair at Ecole Polytechnique has developed computer codes to predict aerodynamic loads and performance of a Darrieus rotor. These models were based on the momentum theory where the rotor is replaced by an actuator disk and then the flow through the rotor is determined by equating the forces on the rotor blades to the time rate of change in momentum through the rotor. Other models include viscous effects, stochastic wind, and a fundamental study of the dynamic-stall phenomenon.

Fig. 2 VAWT and HAWT in Quebec [6]

A. Aerodynamic Codes Based on the DMS Model

Paraschivoiu [10], [11] developed an analytical model, the double-multiple stream tube model (DMS), that considers a multiple-stream tube system divided into two parts where the upwind and downwind components of the induced velocities at each level of the rotor are calculated by using the principle of two actuator disks in tandem, Fig 2.

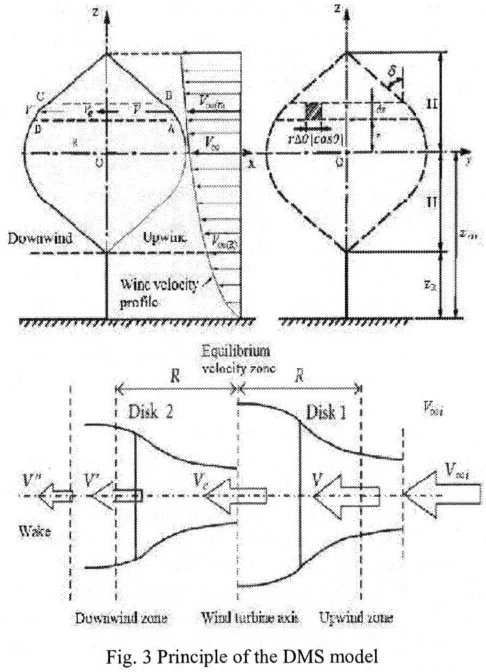

Fig. 3 Principle of the DMS model

For the upstream half-cycle of the rotor, the relative velocity and the local angle of attack as a function of the tips peed ratio "X" are given by:

$$W^2 = V^2[(X - s\quad)^2 + \cos^2\theta \quad s^2\delta] \tag{1}$$

$$\alpha = \arcsin\left[\frac{c}{\sqrt{(X-s\quad)^2+c\quad{}^2\theta\quad s^2\delta}}\right] \tag{2}$$

where $X = \frac{\omega}{V}$ and ω is the turbine rotational speed.

The non-dimensional normal and tangential forces as a function of the azimuthal angle θ are:

$$F_N\ (\theta) = \frac{c}{S}\int_{Z_R}^{Z_R+2H}\left(\frac{W}{V_\infty}\right)^2 C_N d \tag{3}$$

$$F_T(\theta) = \frac{c}{S}\int_{Z_R}^{Z_R+2H}\left(\frac{W}{V_\infty}\right)^2 \left(\frac{C_T}{c}\right) d \tag{4}$$

where c is the local airfoil chord and S is the plane area enclosed by the blade. The normal and tangential force coefficients of the blade section are given by

$$C_N = C_L c \quad + C_D s \tag{5}$$

$$C_T = C_L s\ \alpha - C_D c \tag{6}$$

The lift and drag coefficients C_L and C_D are computed by interpolating the available test data using both the local Reynolds number and the local angle of attack.

Three computer code variants based of the DMS model have been developed: CARDAA, CARDAAV and CARDAAX. CARDAA considers two constant interference factors in the induced velocities for each half-stream tube while CARDAAV code considers the variation of the interference factors as a function of the azimuthal angle on the upwind and downwind half of the rotor. In CARDAAX, the effect of the stream tube expansion is added to the code. Finally, CARDAAS 1D/3D includes a stochastic wind model incorporating the fluctuating nature of the wind. The velocity field of the wind is assumed to be a linear superposition of a mean and a fluctuating components. These codes have been constantly improved in order to produce an efficient software package appropriate for the needs of VAWT designers and have been widely used by many research centers such as Hydro-Quebec (Canada), Sandia National Laboratories (US), and many others [12]. A unique feature of the CARDAAV code, apart from the two actuator disks model, is that additional parameters options can be incorporated into the code, for example the influence of the blade geometry, the airfoil type, the rotating tower, the presence of struts, or the size of the spoiler. Among the principal operating parameters that are readily modifiable to meet the needs of a specific analysis are the wind speed, the rotation speed of the rotor, the local gravitational acceleration and the working fluid properties. Either constant rotation speed for different wind speeds or different rotation speeds for a constant wind speed can be considered when performing an analysis. A power law type variation of the wind speed (as a function of altitude) is taken into account during the computations.

B. Aerodynamic Code Based on Viscous Model

Aerodynamic codes based on the DMS model do not take the viscous effects into account, a computer code named 3DVF [13], [14] has been developed to analyses the Darrieus rotor in a steady incompressible laminar flow field by solving Navier-Stokes equations in a cylindrical coordinates using the finite volume method where the effect of the spinning blades is simulated by distributing source terms in the ring of control volumes that lie in the path of the turbine blades. The calculating method based on the control volume approach uses the widely known "SIMPLER" algorithm. Details of the governing equations with the numerical procedure are given in [15].

C. Dynamic Stall Models

Dynamic stall has a significant influence on the aerodynamic loads and performance of the wind turbine at low tip-speed ratios. It is an unsteady flow phenomenon which refers to the stalling behavior of an airfoil when the angle of attack is changing with time. Dynamic stall is characterized by dynamic delay of stall to angles significantly beyond the static stall angle as well as by massive recirculating regions moving downstream over the airfoil surface. When the Darrieus wind turbine operates at wind speed approaching its maximum, all its blades sections exceed the static stall angle, consequently the whole blade operates under dynamic stall conditions which causes structural fatigue, and even stall flutter leading to catastrophic failure limiting the performance of the turbine. In such conditions, the dynamic lift and drag characteristics present a hysteresis response which is completely different from the static coefficients. The objective of a dynamic-stall model is to propose a methodology to compute the dynamic characteristics from available experimental static coefficients.

To predict the aerodynamic performances and provide aerodynamic loading information to structural dynamic codes, four dynamic stall models are available in our aerodynamic code CARDAAV, three derived from Gormont's method [16], adaptation of [17], adaptation of [18], the modification of [19], and one derived from the indicial method [20].

The effects of dynamic stall are more complex during the turbine operation, in an experimental study [21] performed in a water channel at a tip speed ratio of 2.14, study shows the emission of a leading and trailing edge vortices interacting with the blade trajectory, Fig. 3. It has been suggested to apply the dynamic stall model only in regions of low turbulence as reported by [18].

Using TKFLOW the simulation of the computed flow structure around an airfoil in Darrieus motion based on Navier-Stokes solver in stream function-vorticity formulation has been developed [22]. Since 3-D simulation would be very expensive a 2-D simulation has been adopted. The code has the capability to predict the region where the dynamic stall may occur. Computational Fluid Dynamics (CFD) methods, based on the solution of the Navier-Stokes Equations, are the most accurate (qualitatively and quantitatively) in terms of flow analysis, but their major drawback is the very high computational cost (require large computational resources and very long time to provide converged solutions).

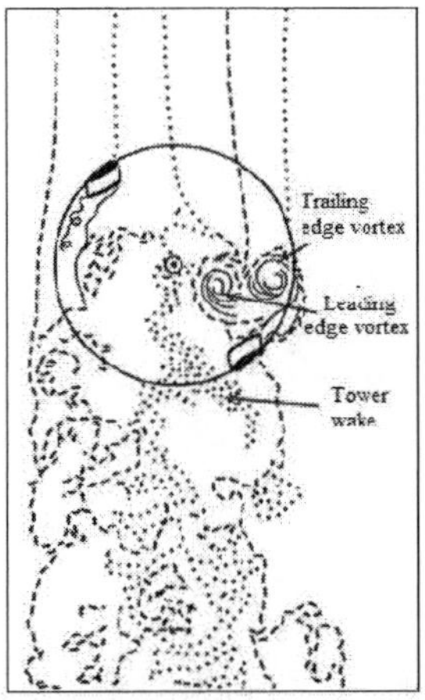

Fig. 4 Flow visualization in the dynamic stall condition for $\lambda = 2.14$

IV. VAWT Design Optimization

For the optimization of the Darrieus wind turbine given a set of constraints, the CARDAAV code has been coupled with a generic algorithm, GENIAL Code, developed by [23]. The variables considered in this optimization are the geometrical and operational parameters such as: height, aspect ratio, chord length, number of blades, airfoils, rotation speed, etc. The use of a genetic algorithm is not only appropriate to analyze a combination of these variables but also to be used for parametric studies since a large number of evaluations are possible due to the high performance of CARDAAV aerodynamic code in term of execution time.

Regarding the use of airfoil section types, the successful design of an efficient rotor can be obtained only when appropriate airfoil sections have been selected. Most VAWTs currently operating worldwide use blades of symmetrical NACA airfoil series. It has been proposed to use Natural Laminar Flow (NLF) airfoils to increase the efficiency of the VAWT [24]. NLF airfoils typically can maintain a laminar flow up to 50% chord through favorable pressure gradients, which results in a significant reduction of the airfoil drag.

A major concern for wind turbine developers is how their aerodynamic designs would fair in rated and extreme weather conditions in terms of structural viability and fatigue. In order to cater to these demands, much effort has been devoted towards the development of static and dynamic structural response of the VAWTs. Newly developed capabilities include a combined aero-structural design of the turbines with focus on structural dynamic response which has been successfully applied also for water turbines [25]. In this regards, Figs. 5-7, show the different steps in the structural dynamic response analysis of a simple 2-bladed H-Darrieus VAWT that include:

a) development of a 3D beam-element model for finite-element analysis (Fig. 5),
b) model analysis (Fig. 6), and
c) the translational and rotational displacement time history (Fig. 7).

The analysis helps identify critical response frequencies as well as help estimate turbine life expectancy under rated and extreme weather conditions.

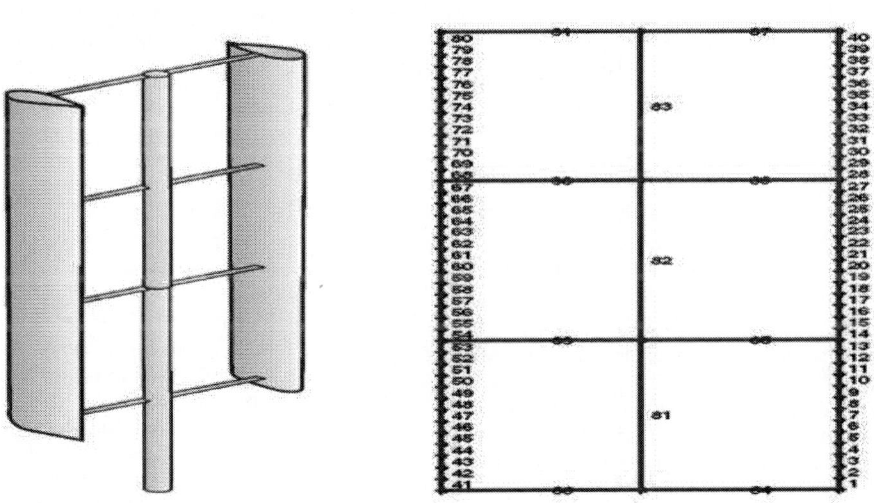

Fig. 5 The H-Darrrieus configuration and the corresponding 3D beam-element model

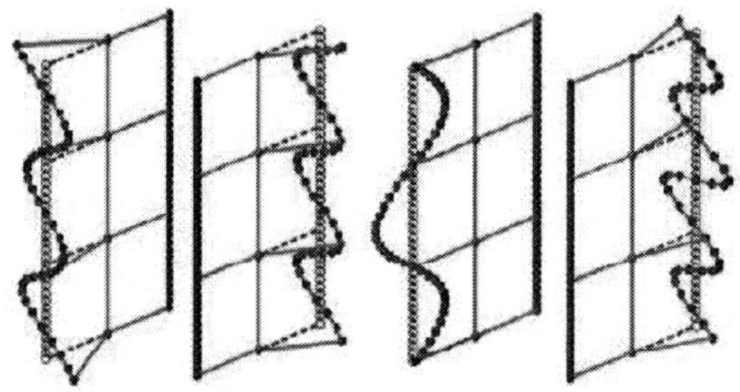

Fig. 6 Different mode shapes for the spinning H-Darrieus turbine

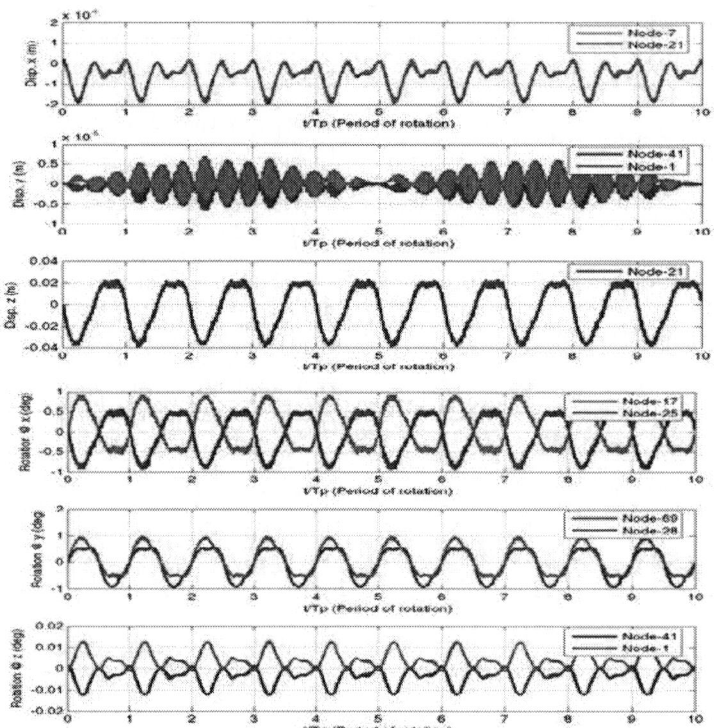

Fig. 7 Time history of maximum translational and rotational displacements of the nodes

V. Results and Discussion

The different capabilities discussed in previous sections make CARDAAV aerodynamic code a very attractive and efficient design and analysis tool. The performance prediction of Sandia 17-m wind turbine, Fig. 8, using CARDAA code was improved by considering the variable interference factors in CARDAAV code and by including stream tube expansion effects in CARDAAX code. We can see from the figure that CARDAA over predicts the power coefficient peak while CARDAAV and CARDAAX results are in good agreements compared to experimental data. Results were also compared to Strickland et *al.* vortex model (VDART3) [26]. The simulation of the dynamic stall hysteresis loop using 3DVF code with indicial model is given by Fig. 9. Compared to CARDAAV results, predictions in the upwind and downwind regions of the turbine are well compared to experimental data [27]. The performance predictions at 42.2 rpm using 3D viscous model it is also well predicted, Fig. 10. In Fig. 11 comparison of the normal force coefficients for the steady state and stochastic loads using CARDAAS 3D shows good agreement with experimental data. Fig. 12 shows the performance of Sandia-17 m machine at 42.2 rpm using the various implementations of the Gormont dynamic stall model where A_{M} is an empirical constant used to correct C_l and C_d for dynamic stall effects [24]. Results are in good agreement with experimental data.

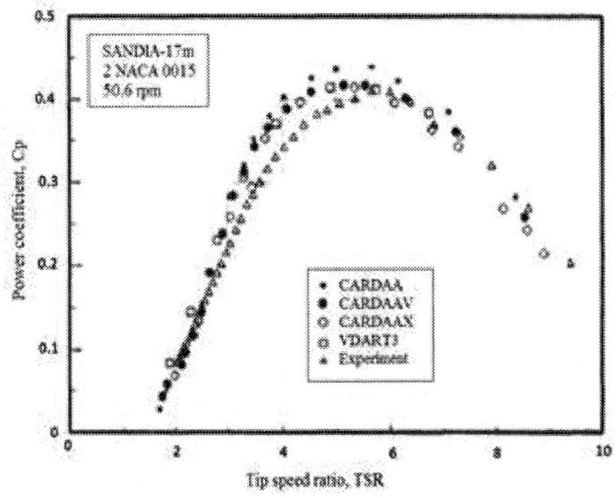

Fig. 8 Power coefficient as a function of the tip-speed ratio, DMS models, Vortex model, and experimental data

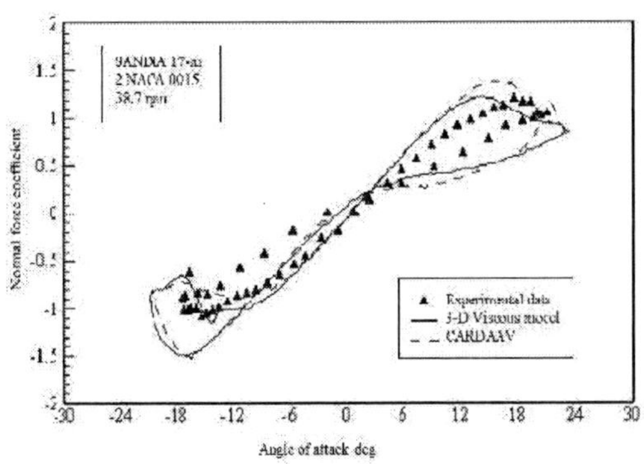

Fig. 9 Normal force coefficient vs. angle of attack at tip-speed ratio of 2.49

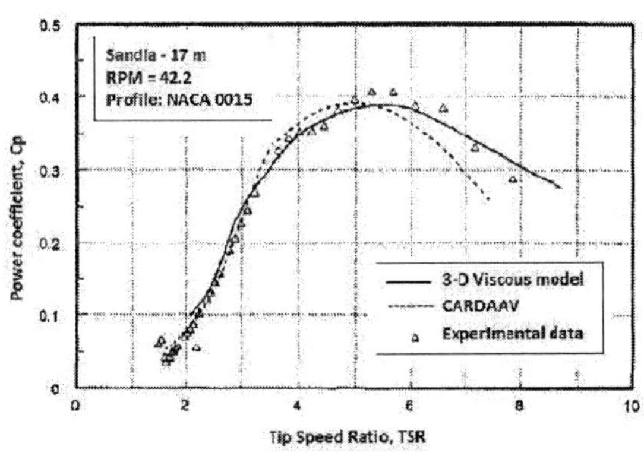

Fig. 10 Power coefficient as a function of TSR, comparison: 3-D viscous and models, CARDAAV, and experimental data

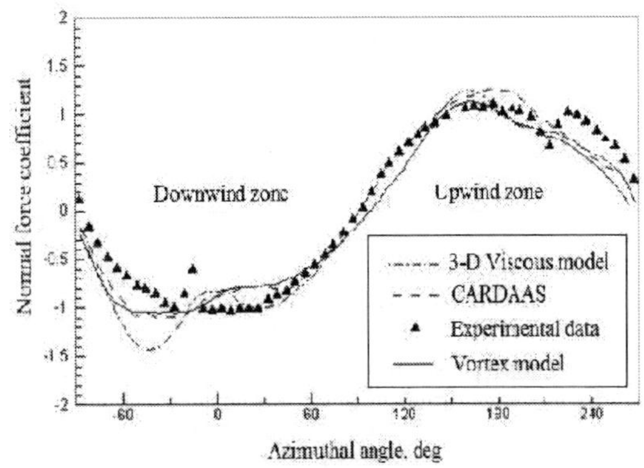

Fig. 11 Comparison of normal force coefficient at tip speed ratio of 2.86 and turbulence = (27%, 25%)

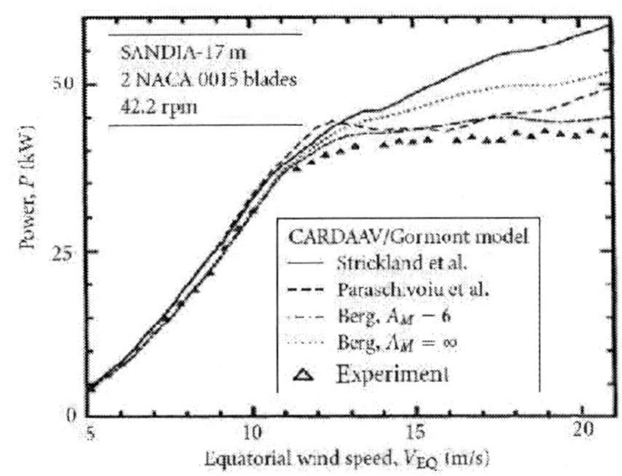

Fig. 12 Power coefficient with Gormont-model adaptations

VI. Conclusion

The Vertical Axis Wind Turbine (VAWT) offers a mechanically and structurally simple method of harnessing the energy of the wind. This simplicity, however, does not extend to the rotor aerodynamics. Aerodynamic loads and performance of the wind turbine depend on the flow field of the wind through the surface swept by the blades. The J. A. Bombardier Aeronautical Chair Group at Ecole Polytechnique de Montreal [28] has conducted many research on the development of computer codes for studying Darrieus rotor aerodynamics, today various aerodynamic code variants based on the double-multiple stream tube model, stochastic wind and viscous flow field, including dynamic stall, are available to designers [1]. Comparison of the analytical results with available experimental data have shown good agreement. The different codes developed represent valuable tools for simulating the flow field around the turbine and dynamic-stall phenomenon. CARDAAV proves to be an efficient and flexible software package, appropriate for the needs of VAWT designers. The capabilities that exist in the aerodynamic codes make CARDAAV a very attractive and efficient design and analysis tool. It computes the aerodynamic loads and rotor performance for VAWTs of any geometry at given operational conditions. It is also possible to couple CARDAAV with an optimization or structural analysis code in order to perform the optimized design of a VAWT.

ACKNOWLEDGMENT

The authors gratefully acknowledge the support of J.-Armand Bombardier Aeronautical Chair.

REFERENCES

[1] Paraschivoiu, I., *Wind Turbine Design with Emphasis on the Darrieus Concept*, Polytechnique International Press, Ecole Polytechnique de Montréal, Montréal, Canada, June 2002.

[2] Tchon, K.F., "Simulation numérique du décrochage dynamique sur un profil d'aile en mouvement de rotation Darrieus," Thèse de doctorat, Dépt. de Génie Mécanique, École Polytechnique de Montréal, 1993.

[3] Rajagopalan, R.G., "Viscous Flow Field Analysis of a Vertical-Axis Wind Turbine," Proceedings of the 21st Intersociety Energy Conversion Engineering Conference

[4] Canadian Wind Energy Association (CanWEA), "Wind Vision 2025, Powering Canada's Future", retrieved September 2015 from http://canwea.ca/pdf/windvision/Windvision_summary_e.pdf

[5] Global Wind Energy Council, "Global Wind Report", Annual Market Update, GWEC, 2014

[6] Canadian Wind Energy Association, (CanWEA), "Installed Capacity", retrieved September 2015 from http://canwea.ca/wind-energy/installed-capacity/

[7] American Wind Energy Association, AWEA, "Wind Energy Facts at a Glance", retrieved September 2015 from http://www.awea.org/

[8] European Wind Energy, (EWEA), "Aiming High, Rewarding Ambition in Wind Energy", retrieved December 2015 from http://www.ewea.org/.

[9] Manwell, J.F., McGowan, J.G., Rogers, A. L, "*Wind Energy Explained, Theory, Design and Application*", John Wiley and Sons, 2009.

[10] Paraschivoiu, I., "Aerodynamic Loads and Performance of the Darrieus Rotor," J. of Energy, Vol., No., pp. 406-412, 1981.

[11] Paraschivoiu, I., and Delclaux, F., "Double- Multiple Streamtube Model with Recent Improvements," *Journal of Energy*, Vol. 7, No. 3, pp. 250–255, May-June 1983.

[12] IONPARA Inc., "Wind Turbines & Aeronautics Consulting", retrieved September 201 from http://www.iopara.ca/clients.shtml.

[13] Paraschivoiu, I., Masson, C., Rajagopalan, R.G., "Predictions and Experiments of the VAWT Viscous Flow Field," AIAA Paper 87-1429.

[14] Allet, A., Paraschivoiu, I., "Numerical Simulation of Three-Dimensional Flow Fields on Vertical-Axis Wind Turbines", The 4r International Symposium on Transport Phenomena and Dynamics of Rotating Machinery, (ISROMAC-4), Hawaii, pp. 214-223, 1992.

[15] Allet, A., "Modele tridimensionnel pour le calcul arodynamique des turbines axe vertical", PhD Thesis, Mechanical Eng. Department, Ecole Polytechnique de Montreal, 1993.

[16] Gormont, R.E., "A Mathematical Model of Unsteady Aerodynamics and Radial Flow for Application to Helicopter Rotors", Technical Report 72-67; NTIS: Springfield, VA, USA, 1973.

[17] Strickland, J.H. Webster, B.T. and Nguyen, T., "A Vortex Model of the Darrieus Turbine: An Analytical and Experimental Study", SAND79-7058, Sandia National Laboratories, Albuquerque, 1980.

[18] Paraschivoiu, I. "Double-Multiple Stream Tube Model for Studying Vertical-Axis Wind Turbines", Journal of Propulsion and Power, Vol. 4, pp. 370-378, 1988.

[19] Berg, D.E. An Improved Double-Multiple Stream Tube Model for the Darrieus-Type Vertical Axis Wind Turbine, Sixth Biennal Wind Energy Conference and Workshop, pp. 231-233, 1983.

[20] Beddoes, T.S., and Leishman, J.G., "A Semi-Empirical Model for Dynamic Stall", Journal of the American Helicopter Society, Vol. 34, no. 3, pp. 3-17, 1989.

[21] Brochier, G.; Fraunié, P.; Béguier, C.; Paraschivoiu, I. Water Channel Experiments of Dynamic Stall on Darrieus Wind Turbine Blades. AIAA Journal of Propulsion and Power 1986, Vol-2, pp. 445–449.

[22] Tchon, K.F., "Simulation numérique du décrochage dynamique sur un profil d'aile en mouvement de rotation Darrieus," Thèse de doctorat, Dépt. de Génie Mécanique, École Polytechnique de Montréal, 1993.

[23] Widell, H., "Genial 1.1, a Friendly Function Optimizer based on Evolutionary Algorithms," 1997.

[24] Masson, C. Leclerc, C., and Paraschivoiu, I., "Appropriate Dynamic-Stall Models for Performance Predictions of VAWTs with NLF Blades", International Journal of Rotating Machinery 1998, Vol. 4, No. 2, pp. 129-139.

[25] Norbert V. Dy, Farooq Saeed and Ion Paraschivoiu, "Dynamic Response Analysis of Darrieus-Type Vertical Axis Water Turbines", Proceedings of the Twenty-third International Offshore and Polar Engineering Anchorage, Alaska, June 30–July 5, 2013. The International Society of Offshore and Polar Engineers (ISOPE).

[26] Strickland, J.H., Webster, B.T., and Nguyen, T., "A Vortex Model of the Darrieus Turbine: An Analytical and Experimental Study," Sandia National Laboratories, Report, SAND79-7058, 1980.

[27] Akins, R. E., "Measurements of Surface Pressure on an Operating Vertical Axis Wind Turbine", Sandia National Laboratories SAND89-7051, Albuquerque, NM, 1989.

[28] J.-A. Bombardier Aeronautical Chair, Paraschivoiu, I., Ecole Polytechnique de Montreal, retrieved September 2015 from http://www.polymtl.ca/recherche/rc/en/professeurs/details.php?NoProf=229.

World Academy of Science, Engineering and Technology
International Journal of Mechanical, Aerospace, Industrial, Mechatronic and Manufacturing Engineering Vol:10, No:12, 2016

Numerical Study of Flapping-Wing Flight of Hummingbird Hawkmoth during Hovering: Longitudinal Dynamics

Yao Jie, Yeo Khoon Seng

International Science Index, Aerospace and Mechanical Engineering Vol:10, No:12, 2016 waset.org/Publication/10005996

***Abstract*—**In recent decades, flapping wing aerodynamics has attracted great interest. Understanding the physics of biological flyers such as birds and insects can help improve the performance of micro air vehicles. The present research focuses on the aerodynamics of insect-like flapping wing flight with the approach of numerical computation. Insect model of hawkmoth is adopted in the numerical study with rigid wing assumption currently. The numerical model integrates the computational fluid dynamics of the flow and active control of wing kinematics to achieve stable flight. The computation grid is a hybrid consisting of background Cartesian nodes and clouds of mesh-free grids around immersed boundaries. The generalized finite difference method is used in conjunction with single value decomposition (SVD-GFD) in computational fluid dynamics solver to study the dynamics of a free hovering hummingbird hawkmoth. The longitudinal dynamics of the hovering flight is governed by three control parameters, i.e., wing plane angle, mean positional angle and wing beating frequency. In present work, a PID controller works out the appropriate control parameters with the insect motion as input. The controller is adjusted to acquire desired maneuvering of the insect flight. The numerical scheme in present study is proven to be accurate and stable to simulate the flight of the hummingbird hawkmoth, which has relatively high Reynolds number. The PID controller is responsive to provide feedback to the wing kinematics during the hovering flight. The simulated hovering flight agrees well with the real insect flight. The present numerical study offers a promising route to investigate the free flight aerodynamics of insects, which could overcome some of the limitations of experiments.

***Keywords*—**Aerodynamics, flight control, computational fluid dynamics, flapping-wing flight.

I. Introduction

INSECTS represent some of the most versatile and manoeuvrable flying animals. Many of them can hover, take off backwards, decelerate rapidly, and some insects can even land upside down. This kind of rapid energy-efficient flight has permitted access to more abundant ecological resources and rapid escape from predators. The remarkable capacity for mobility and manoeuvrability in air has greatly interested biologists, scientists and engineers to study and understand the physical phenomenon of insect flight, which can play key role in improving Micro Air Vehicle (MAV) design. Yet, insects cannot generate enough lift to fly in the sense of conventional laws of aerodynamics, which rely on steady state approximations. The extra lift required must be generated from the complex (unsteady) flapping motions generated during the wing-flapping cycle [1]. Until recently, the exact nature of this extra source of lift remains a mystery. By understanding the aerodynamics associated with insect flight, a new lift generation system can be developed, which would be a great breakthrough in aerodynamics.

Experimental biomechanics and fluid dynamics have played a key role in the study of flapping flight for a long time. Experiment has its limitation in understanding the physics of flapping flight due to the difficulty in observing wing kinematics with small physical scales, visualizing unsteady flow with great complexity and controlling test subjects with desired motion [2], [3]. The Computational Fluid Dynamics (CFD) has offered a more convenient and affordable approach for the unsteady aerodynamic study. Benefiting from the continuously improving computing hardware and technology, the development of CFD has advanced rapidly in the past several decades. A diverse variety of numerical schemes have been developed to assist in the study of aerodynamics in aviation, automobile, marine engineering, environmental engineering and many other industrial and research areas [4], [5].

CFD tools are widely available in standard software packages such as Fluent, CFX, Star-CD, etc. now for both industrial and scientific applications. These standard software packages show reliability and convenience in common fluid flow problems especially for steady flow problems. The complex kinematics of flapping wing motion makes it necessary to use moving mesh or other re-meshing methods to resolve mesh deformation problem [6]. In addition, the rigid body motion has to be solved together with the flow field when controlled flight is involved. Since most of the software packages have adopted only Eularian methods, they are not suitable for immersed boundary problems, especially when the boundary has complicated motion. This indicates that current standard software packages have inherent drawbacks for solving flapping flight problems, such as poor numerical accuracy and high computation cost.

Singular Value Decomposition based Generalised Finite Difference (SVD-GFD) is a novel numerical scheme that was continually developed in the Department of Mechanical Engineering, National University of Singapore. This numerical scheme is capable of achieving high numerical accuracy in solving for complex immersed boundary (IB) problems with

Yao Jie, Yeo Khoon Seng are with the National University of Singapore, 119077, Singapore (e-mail: yaojie@u.nus.edu).

improved computational efficiency as compared to the generally adopted Finite Element and Finite Volume methods in solving such problems [7]-[9].

The purpose of the current research work is to use the aforementioned CFD scheme to simulate and study insect flapping flight problems and attempt to gain understanding of the unsteady aerodynamics involved in flapping wing flights and maneuvering.

II. Methodology

Insect models are used in current numerical simulations. Hummingbird hawkmoth (*Macroglossum stellatarum*) is adopted as the model flapping wing flyer with relatively high Reynolds number (Fig. 1).

Fig. 1 Hummingbird hawkmoth (*Macroglossum stellatarum*)

The flapping wing motion of hummingbird hawkmoth has a frequency f of approximately 70 Hz and a stroke amplitude Φ of about 115° according to literature [10]. The Reynolds number of the flyer with a reference length of wing mean chord $\overline{c}$ and a reference velocity of wing tip U_{ref} is

$$\mathrm{Re} = \frac{\overline{c}U_{ref}}{\nu} = \frac{\overline{c}\cdot R\cdot 2\Phi\cdot f}{\nu} = \frac{4R^2\Phi f}{\nu\cdot AR} \quad (1)$$

where $\nu = 1.5\times10^{-5}$ m^2·s^{-1} is the kinematic viscosity of air, $AR = (2R)^2/S$ is the aspect ratio with the surface area of a wing pair $S = 2R\overline{c}$. The Reynolds number thus works out to be around 3000, which is much higher than that of fruit fly (148) [6].

The flyer's geometric model is shown in Fig. 2 with global coordinate (x, y, z) and insect body coordinate $(x, y, z)^b$. The insect model is created with a pre-set pitch angle of 45° in the body coordinate system; because a hummingbird hawk moth hovers with body pitch approximately that angle. The body coordinate is aligned with the global coordinate initially, which is also the hovering equilibrium position.

The governing equation of fluid flow in insect flapping flight problems is the three-dimensional incompressible Navier-Stokes equations, given in the non-dimensional Arbitrary Lagrangian-Eulerian (ALE) form:

$$\nabla\cdot\mathbf{u} = 0 \quad (2)$$

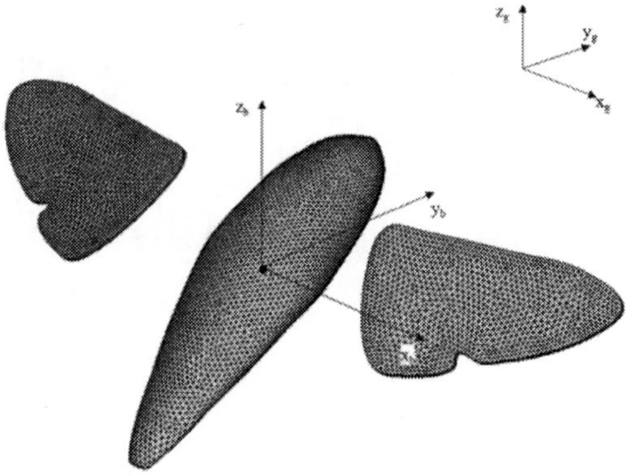

(a)

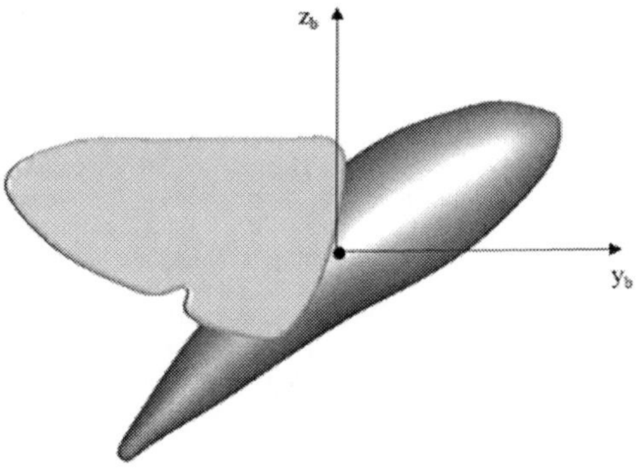

(b)

Fig. 2 Drawing of insect model and coordinate systems: (a) Perspective view with surface mesh. (b) Side view. The longitudinal axis of insect body is selected as the y_b axis of body frame. The insect model is created with a pre-set pitch angle of 45° in the body frame

$$\partial_t\mathbf{u} = -(\mathbf{u}-\mathbf{u}^g)\cdot\nabla\mathbf{u} + \frac{1}{\mathrm{Re}}\Delta\,\mathbf{u} - \nabla p \quad (3)$$

where $\mathbf{u}(\mathbf{x}, t)$ and $p(\mathbf{x}, t)$ represent the velocity and pressure fields of the fluid domain respectively. The $\mathbf{u}^g$ is the convection velocity of the computational node, which is equal to zero for stationary nodes.

The governing equations are solved on a hybrid background Cartesian grid nodes and clouds of meshfree grids around IBs. The standard 7-point central finite difference scheme is applied at Cartesian nodes that do not have meshfree nodes nearby, while the SVD-GFD scheme is applied at the meshfree nodes and Cartesian nodes with meshfree nodes nearby. A second-order implicit projection method, based on a fractional-step Crank-Nicolson scheme is applied here to solve the ALE form NS equations. Detail of this method has been discussed by [7] and [9], which will not be elaborated here.

The flyer is located in a cubic domain with boundaries set at a distance of about three wing lengths away from the flyer. The grid is non-uniform with coarse grid on the outside and fine grid near the centre where the model insect is located to give good resolution to the immediate flow around the flyer.

Considering the flapping wing flyer with mass $\boldsymbol{M}$ as one rigid body, and denoting the linear and angular velocities at its centre of mass $\boldsymbol{C}$ as $\mathbf{V}_C(t)$ and $\boldsymbol{\omega}(t)$ respectively, the governing dynamical equations for the flyer are then given by

$$\begin{aligned}
&\frac{\mathrm{d}\mathbf{X}_C(t)}{\mathrm{d}t} = \mathbf{V}_C(t) \\
&\frac{\mathrm{d}\boldsymbol{\Theta}(t)}{\mathrm{d}t} = [\mathbf{K}(\boldsymbol{\Theta})] \cdot \boldsymbol{\omega}(t) \\
&\frac{\mathrm{d}(M \cdot \mathbf{V}_C)}{\mathrm{d}t} = -Mg\mathbf{i}_z + \int_{\Gamma(t)} [\boldsymbol{\sigma}] \cdot \mathbf{n} d\Gamma \\
&\frac{\mathrm{d}([\mathbf{I}(t)] \cdot \boldsymbol{\omega}(t))}{\mathrm{d}t} = \int_{\Gamma(t)} (\mathbf{x} - \mathbf{X}_C(t)) \times ([\boldsymbol{\sigma}] \cdot \mathbf{n}) d\Gamma
\end{aligned} \quad (4)$$

$\mathbf{X}_C(t)$ is the location of CoM at time t. $\boldsymbol{\Theta}(t)$ is the rotation angle vector of the flyer. $[\mathbf{K}(\boldsymbol{\Theta})]$ is a transformation matrix, which transforms the angular velocity $\boldsymbol{\omega}(t)$ of the flyer into the rate of change of its orientation vector $\boldsymbol{\Theta}(t)$. $[\boldsymbol{\sigma}]$ is the Newtonian fluid stress tensor which can be obtained from CFD solver. $[\mathbf{I}(t)]$ is the inertia tensor of the flyer, Γ denotes the surface of the rigid body and $\mathbf{x}$ denotes the location of an arbitrary point on the surface. More details about the fluid dynamics and fluid-body interaction involved in the research can be found in [6].

III. Flight Control

An insect has 6 degrees of freedom in total when it flies; they are surge (forward/backward), heave (up/down), sway (left/right), roll, yaw and pitch. Surge, heave and pitch are grouped together as longitudinal motion, and the remaining three are grouped as lateral motion. Since the longitudinal motion and lateral motion are usually coupled weakly, they can be analysed separately. In this paper, only the longitudinal flight is investigated.

The stroke-plane adjustment can be used to control the trust forces and pitch moment generated by the wing pair, and thus to control the longitudinal motion. The adjustments of stroke-plane are done by rotation the wing frame around the affixed wing roots. Small pitching rotation of the stroke plane can generate horizontal forces (y-direction) to control the forward/reverse motion of the flyer (Fig. 3). For the hovering flight, the disturbance in the mean lift force and mean thrust are generally quite small. Hence, relatively small rotations of the stroke plane are needed to neutralize or counter them. This means that subtle modulation of the wingbeat frequency f will be normally be sufficient to moderate changes in the net vertical force during hovering. Biasing the stroking action of the wings in the stroke plane can shift the cycle-averaged centre of the wing force, and generate desired moment to counter the pitching disturbance of the flyer. The described stroke-plane adjustment thus comprises three independent controlling actions: pitching rotation of stroke plane (β), biasing the mean positional angle of the wing stroke (γ) and wingbeat frequency (f) modulation. The orientation of wing frame and wingbeat frequency is updated at the beginning of each wingbeat. If the orientation of wing frame is changed, sixth order polynomials will be used to describe the rotation of wing frame from old orientation to the new one within one wingbeat.

A generic PID based feedback control algorithm is used in the present study. All the variables are controlled independently in both the stroke-plane adjustments and intra-stroke adjustments. The PID feedback control system comprises three components: a proportional component P that reflects the current deviation (also termed the error) from the desired state, an integral component, I, that takes into account recent error events, and a differential component, D, for estimating likely future error based on the current rate of deviation/error change. The correction or control vector that it works out is a weighted combination of these three components to give

$$\mathbf{u}(t) = K_P\mathbf{e}(t) + K_I\int_0^t \mathbf{e}(\tau)\mathrm{d}\tau + K_D\frac{\mathrm{d}}{\mathrm{d}t}\mathbf{e}(t) + \mathbf{u}(0) \quad (5)$$

More details about the flight control involved in the research can be found in [6].

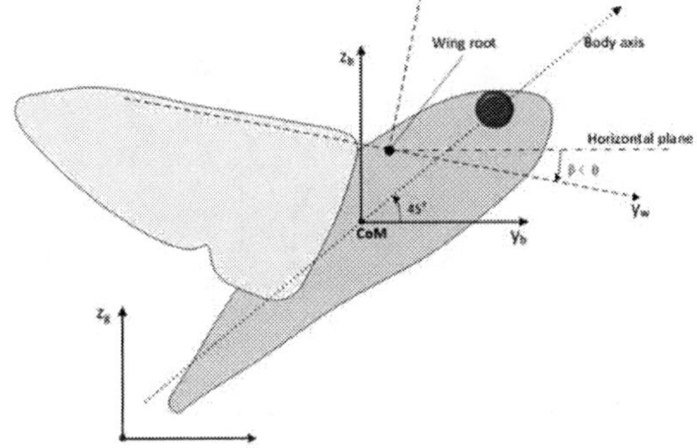

Fig. 3 Orientation of wing frame/stroke-plane: The wing stroke-plane forms an angel β (stroke plane angle) with respect to horizontal when the insect is viewed laterally

IV. Effect of Control Actions

Fixed body simulation with prescribed wing flapping motion is conducted to analyse the effect of different control actions. Mean positional angle (γ), wing plane angle (β) and wing beating frequency (f) are varied to study how the force and torque acting on the insect are changed. The results are summarized in Figs. 4-6. It can be found that the effects of γ and f are unique, the former one varies the pitching torque only and the latter one aims to change the vertical force. However, the wing plane angle (β) can change both the longitudinal force and pitching torque, the latter one of which is undesired. To remove such effect, we need to add one more term to the final mean positional angle, in additional to the PID controller output.

$$\gamma = \gamma_{PID} - k\beta \tag{6}$$

$k \approx 0.3$, which is the ratio of the gradients in Figs. 5 (f) and 4 (b). In such way, each control parameter takes care of one degree of freedom independently, where the insect flight is easier to control.

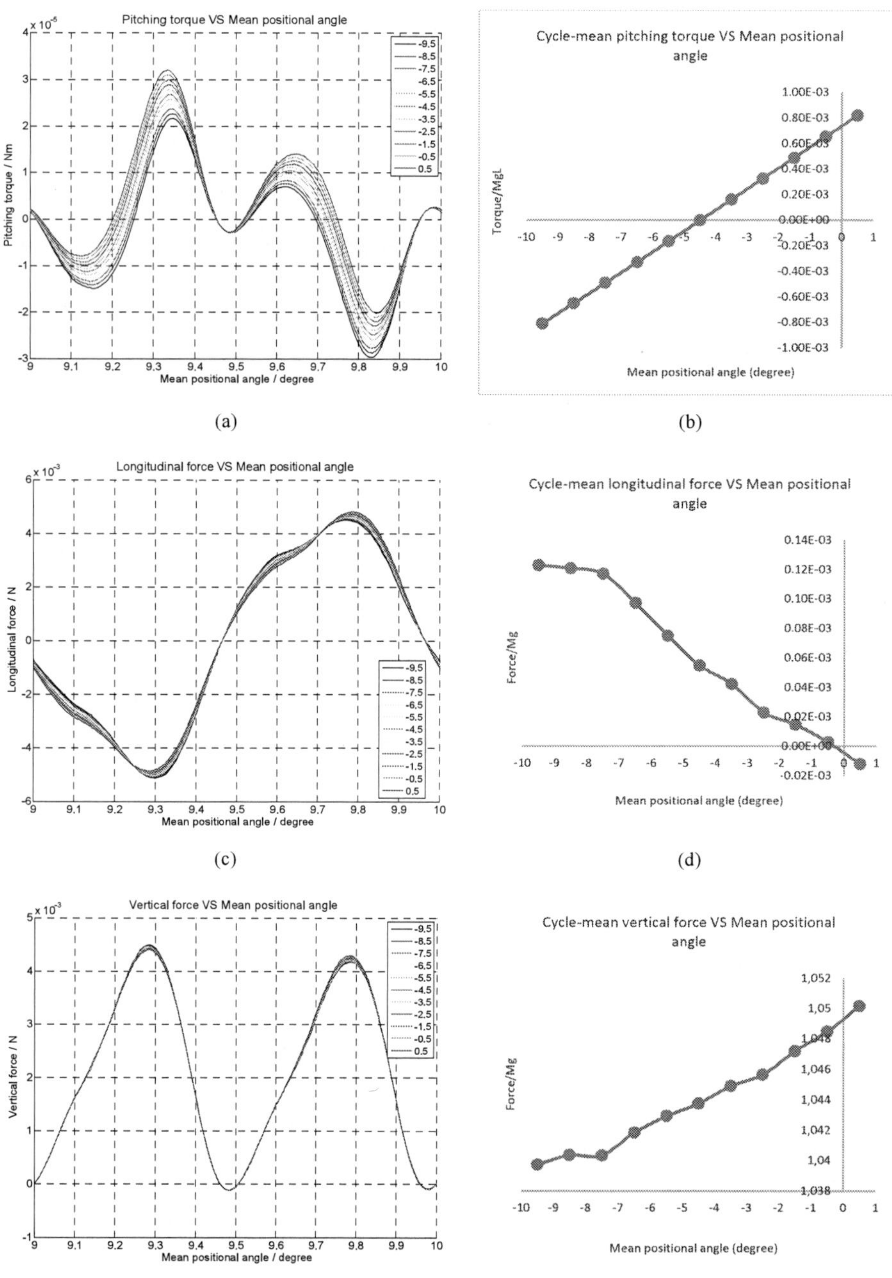

Fig. 4 Effect of mean positional angle: It mainly changes the pitching torque, which varies linearly near the equilibrium position; Change of the vertical force and longitudinal force is within 2%

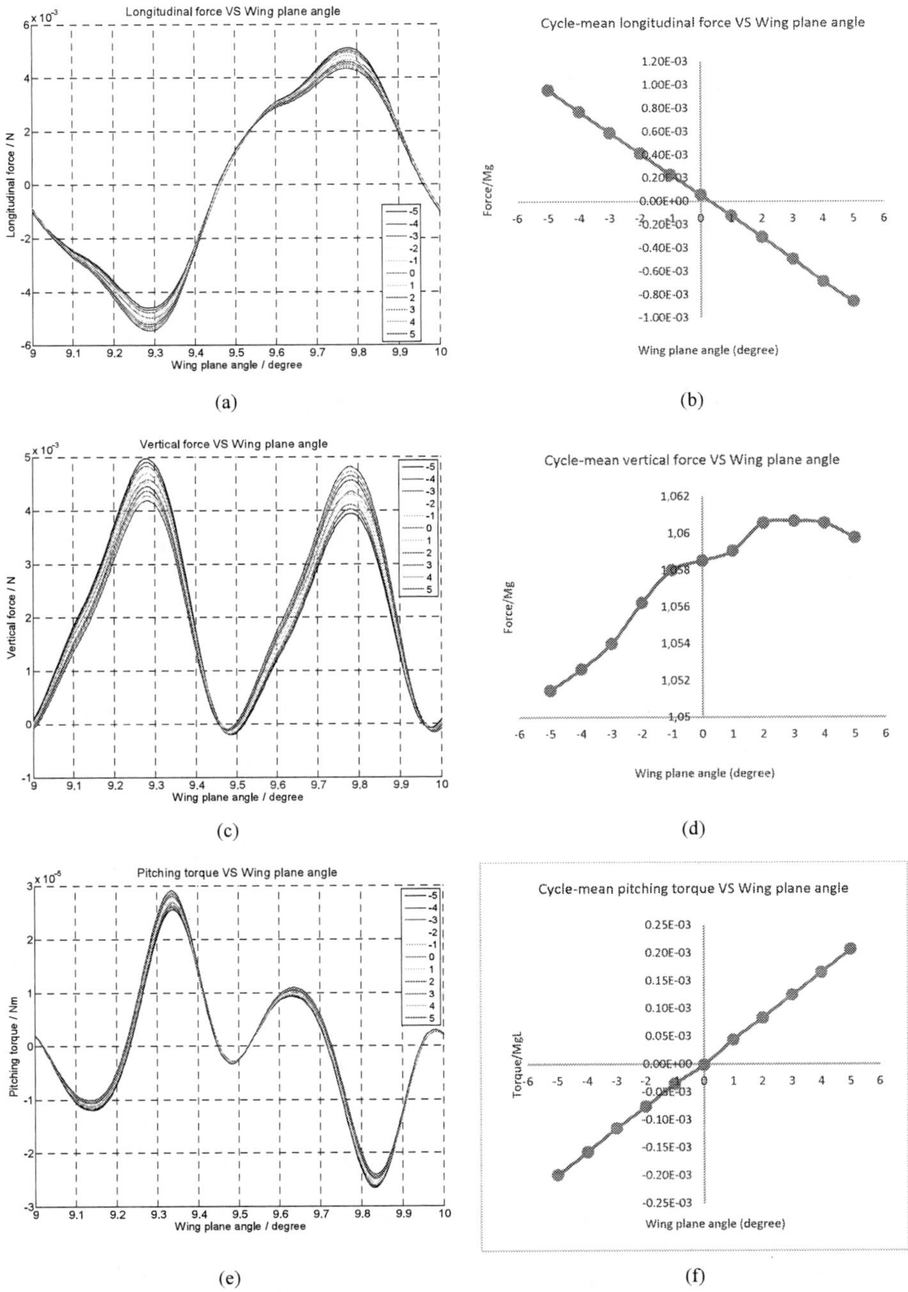

Fig. 5 Effect of wing plane angle: It changes the longitudinal force and pitching torque, both of which vary linearly near the equilibrium position; Change to the vertical force is negligible

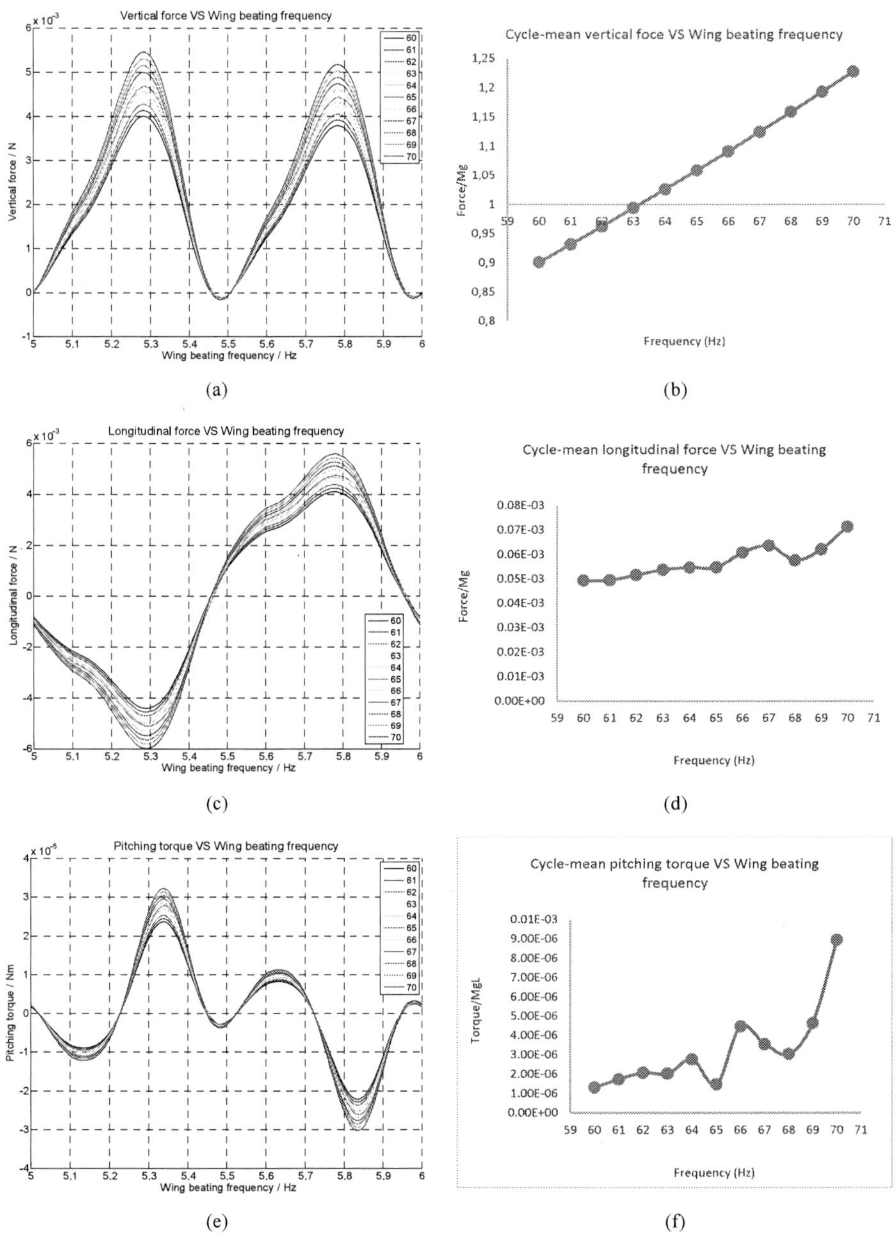

Fig. 6 Effect of wing beating frequency: It mainly changes the vertical force, which varies linearly near the equilibrium position; Change of the pitching torque and longitudinal force is nonsignificant

V. Controlled Hovering Flight

The PID controller is carefully tuned in each degree of freedom, the insect hovering flight can be well controlled.

Figs. 7 and 8 show the hovering behaviour of the model hummingbird hawkmoth over a period of 50 wing cycles. The simulation shows how the body oscillate during hovering, which is an expected phenomenon due to the cyclical nature of the aerodynamic forces and moment. The action of the controller to sustain hovering at the designated position and attitude is shown in Figs. 9 and 10. It can be found that the amplitude of pitch oscillation is controlled within 10 degrees for the overall simulation and is within 5 degrees for one whole oscillation when the flight is stabilized in the end. The pitch oscillation is in a reasonable range compared to the real hummingbird hawkmoth [10]. In addition, the position of the insect is kept in close vicinity of the designated location, to be more specific, within the range of 15 percent of the wing length. Overall, the control coefficients are able to make the insect hover with good performance.

When the insect starts to drift backward and pitch upward at the beginning of the simulation, the controller tries to restore the insect to the designated status by increasing the forward force and pitching down torque. That is done by pitching the stroke plane forward (downward) and rotating the mean positional angle backward. Simultaneously, the insect moves upward slightly at the beginning and it is brought back by the controller by reducing the wing beating frequency, which in turn decreases the lift. It can be deduced from the controller record that the currently used PID controller is very responsive and can make the insect hover well. Furthermore, the average mean positional angle is around -4.5 degree, that is to say, the symmetric axis of the wing motion has to be swept backward by -4.5 degree so that the acting point of the mean lift force is just above the CoM. The wing beating frequency change very little during the hovering, as shown by Figure 10. The reason is that small change in frequency is enough to correct for lift deficiency.

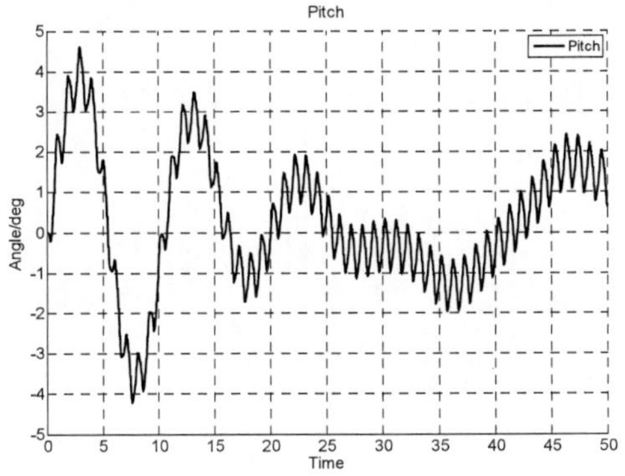

Fig. 7 Evolution of pitch angle during hovering

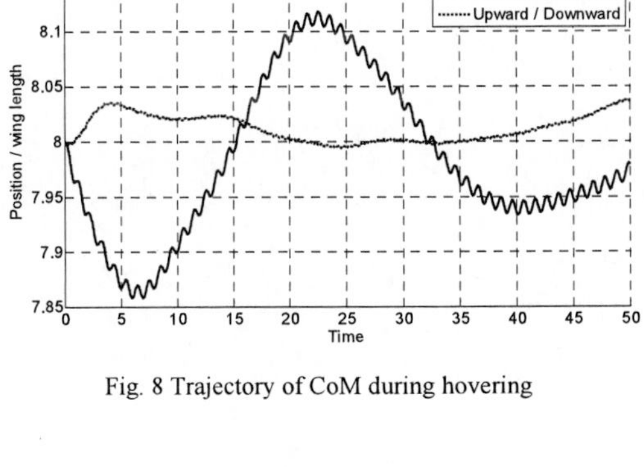

Fig. 8 Trajectory of CoM during hovering

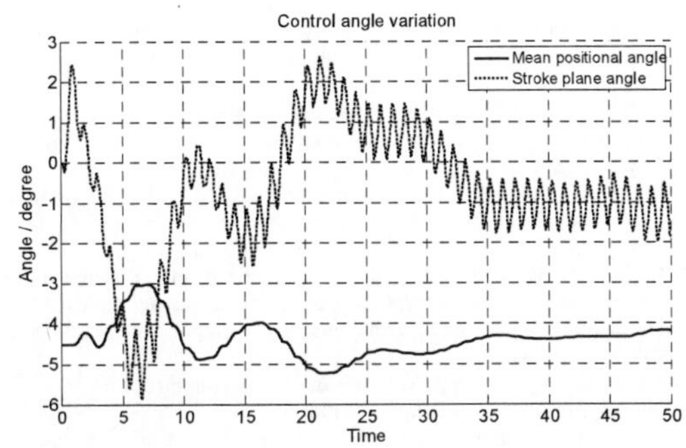

Fig. 9 Variations of stroke plane angle β and mean positional angle γ

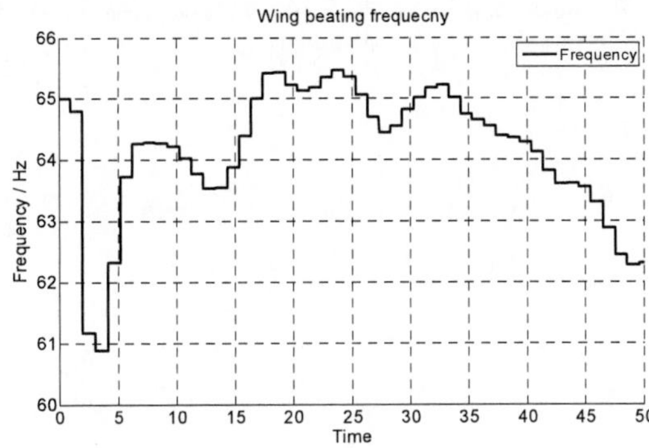

Fig. 10 Evolution of wing beating frequency

VI. Conclusion

In the present study, the SVD-GFD based CFD solver is adapted to study the dynamics of a free hovering flapping-wing flyer. A complex irregular shaped body and wings of the hummingbird hawkmoth has been modelled with rigid wing assumptions. The PID controllers are established to achieve the designated flight. The controllers are proven to be robust and effective at adjust the wing kinematics to stabilize the flyer and control the direction of normal hovering flight. The present modelling approach offers a promising route of investigation that could complement as well as overcome some of the limitations of experiments in the area of free flight aerodynamics of insects. Study shows that the hummingbird hawkmoth is able to hover with reasonable pitch amplitude and very small displacement from the designated location. The PID control mechanism is shown to be stable and responsive to make the hummingbird hawkmoth hover.

References

[1] Ellington, C.P., et al., Leading-edge vortices in insect flight. 1996.
[2] Brodsky, A., Vortex formation in the tethered flight of the peacock butterfly *Inachis io* L. (Lepidoptera, Nymphalidae) and some aspects of insect flight evolution. Journal of experimental biology, 1991. 161(1): p. 77-95.
[3] Lua, K., et al., On the aerodynamic characteristics of hovering rigid and flexible hawkmoth-like wings. Experiments in fluids, 2010. 49(6): p. 1263-1291.

[4] Sun, M. and J. Tang, Unsteady aerodynamic force generation by a model fruit fly wing in flapping motion. Journal of Experimental Biology, 2002. 205(1): p. 55-70.

[5] Miller, L.A. and C.S. Peskin, A computational fluid dynamics of clap and fling' in the smallest insects. Journal of Experimental Biology, 2005. 208(2): p. 195-212.

[6] Wu, D., K. Yeo, and T. Lim, A numerical study on the free hovering flight of a model insect at low Reynolds number. Computers & Fluids, 2014. 103: p. 234-261.

[7] Chew, C.S., K. Yeo, and C. Shu, A generalized finite-difference (GFD) ALE scheme for incompressible flows around moving solid bodies on hybrid meshfree–Cartesian grids. Journal of Computational Physics, 2006. 218(2): p. 510-548.

[8] Wang, X., et al., A SVD-GFD scheme for computing 3D incompressible viscous fluid flows. Computers & Fluids, 2008. 37(6): p. 733-746.

[9] Yu, P., et al., A three-dimensional hybrid meshfree-Cartesian scheme for fluid–body interaction. International Journal for Numerical Methods in Engineering, 2011. 88(4): p. 385-408.

[10] Wu, G. and L. Zeng, Measuring the kinematics of a free-flying hawk-moth (*Macroglossum stellatarum*) by a comb-fringe projection method. Acta Mechanica Sinica, 2010. 26(1): p. 67-71.

Vehicle Aerodynamics: Drag Reduction by Surface Dimples

C. K. Chear, S. S. Dol

***Abstract*—**For a bluff body, dimples behave like roughness elements in stimulating a turbulent boundary layer, leading to delayed flow separation, a smaller wake and lower form drag. This is very different in principle from the application of dimples to streamlined body, where any reduction in drag would be predominantly due to a reduction in skin friction. In the present work, a car model with different dimple geometry is simulated using k-ε turbulence modeling to determine its effect to the aerodynamics performance. Overall, the results show that the application of dimples manages to reduce the drag coefficient of the car model.

***Keywords*—**Aerodynamics, Boundary Layer, Dimple, Drag, Kinetic Energy, Turbulence.

I. Introduction

THE successful application of dimples on a golf ball has brought inspiration to the vehicle aerodynamics engineers. A golf ball with dimple surface can travel higher and further than a smooth surface golf ball when subjected to identical force (i.e. club swing). Dimples on the golf ball induce turbulence at lower Reynolds number (a non-dimensionless number that determines whether the flow is laminar or turbulent) providing extra momentum or energy to the boundary layer causing delay in flow separation, as shown in Fig. 1 [4], [6]. This phenomenon causes smaller wakes area (i.e. swirling flow region) behind the ball thus reducing the total drag [2].

Every single body that moves in fluid will experience drag. Basically, the drag force can be divided into two different types which are pressure drag and friction drag [4], [6]. Friction drag comes from friction between the fluid and the contacted surface. Friction drag is related with the development of boundary layer (thin layer of flows close to the surface due to fluid viscosity) and the Reynolds number. Pressure drag forms due to the eddying motions that are set up in the fluid by the passage of the body. Pressure drag is related with the formation of a wake (i.e. shape or form of the body). When drag force of a body is dominated by friction drag, the body will be defined as a streamline body (i.e. fish, bird, or an airfoil that is in small or zero angles of attack). On the other hand, the drag force of a bluff body (i.e. brick, golf ball or an airfoil that is in large angle of attack) is dominated by pressure drag.

C.K. Chear is with the Department of Mechanical Engineering, Curtin University, Miri, Sarawak, Malaysia.
S. S. Dol is with the Department of Petroleum Engineering, Curtin University, Miri, Sarawak, Malaysia (e-mail: sharulsham@curtin.edu.my).

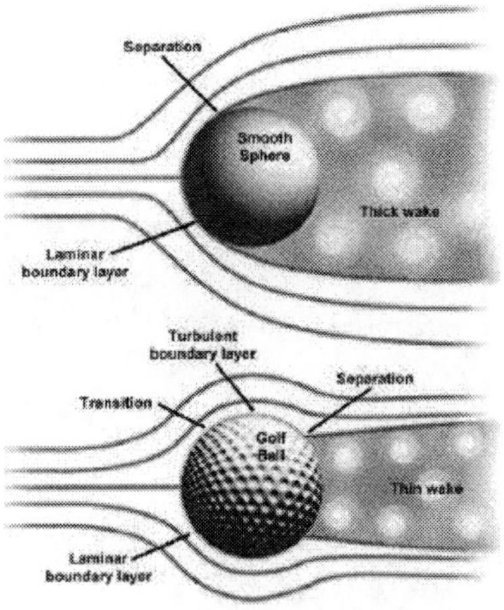

Fig. 1 Dimple pattern on golf ball

In a famous science entertainment television program Mythbusters in 2009 has made an experiment on applied dimples on the car surface (Fig. 2). In the program, Jamie and Adam have created dimples on a car surface with clay. They found out that there was increased of fuel consumption efficiency. The efficiency was increased by 11% when compared to a normal surface car, which was considered as a major improvement. Although it was not an official experiment or research, but it has shown that there is a possibility to apply dimples on the car surface or even other transportation mode like trains and aircrafts to reduce fuel consumption.

Fig. 2 The Mythbusters crew experimented with dimples on this Ford Taurus

Several previous studies (experimental and numerical) observed that dimples on an airfoil create extra turbulence to delay the boundary layer separation. The effect was improved when the airfoil was at angle of attack. For example, [5] has

International Science Index, Mechanical and Mechatronics Engineering Vol:9, No:1, 2015 waset.org/Publication/10000829

suggested a smart dimple matrix over the airfoil which will sense boundary layer separation and arrange dimple in the least drag and high lift configuration, as shown in Fig. 3. German researchers proved an effect which could revolutionize the design of ships, vehicles and aircrafts. Experiments in the wind tunnel showed: dimples in the surface of a body massively reduce the flow resistance. They came to an unexpected result: the train with dimples has 16% less frictional resistance, depending on the speed. The finding seems contradicting the theorem that says surface roughness (dimple) increases the surface area thus friction. Moreover, the drag reduction was overcome by the newly appearing pressure force contribution due to surface unevenness [3].

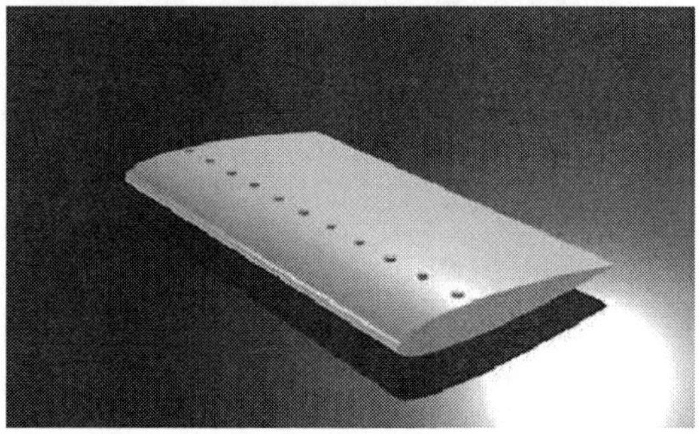

Fig. 3 Dimples matrix as suggested by [5]

So, will the dimples bring a significant effect to streamline body (i.e. vehicle design) as it brings to the bluff body? Adding a dimple on a streamlines body might help to delay the flow separation and reduce the size of the wake but it might be increase the friction drag as a trade-off. As the subject is enormously important to save energy and to protect the global environment, studies to further investigate the effect of dimple on the aerodynamics of the car body have been made with the aim to determine how much changes the dimple will bring to the coefficient of drag.

II. Methodology

A. Car Model

In this work, Ahmed body car model with 25 degree slant angle will be used. Ahmed body car model is a simplify car model for accurate flow simulation but still retain with the significant features of a car which are curved forebody, straight center section and the angled rear end. It is a typical bluff body which commonly using for the simulation to study the flow past a car [1]. With the dimension details from Ahmed research, the model will be drafted by using Autodesk AutoCAD software. After the default model is drafted, it will be added with different ratio of dimple at the near back of the roof. The body geometry is shown in Fig. 4.

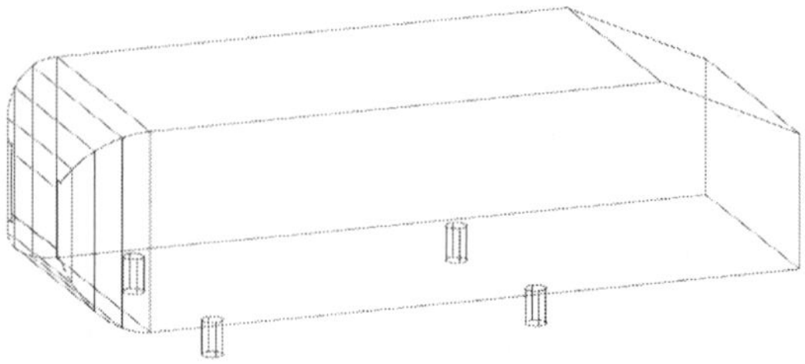

Fig. 4 Isometric view of the Ahmed body car model

B. Dimple Ratio

In order to have more precise result, a parameter called dimple ratio (DR) has been introduced. Basically, dimple ratio is the ratio between the depths of the half dimple over the print diameter of the dimple (Fig. 5). In this work, car model with be simulated with the DR of 0.05, 0.2, 0.3 0.4 and 0.5.

$$\text{Dimple ratio, DR} = \text{depth}, h / \text{diameter}, D \quad (1)$$

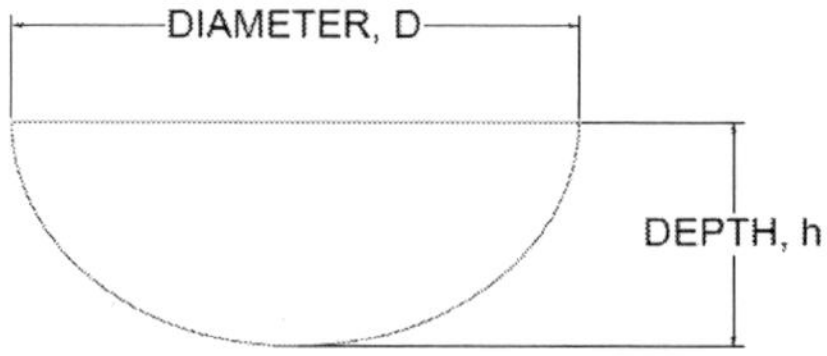

Fig. 5 Dimple ratio

C. Meshing

The meshing is done by using ANSYS meshing software. Tetrahedral meshing is used for the analysis (Fig. 6). Table I shows the setting that has been applied to all models. It shows that the skewness and the orthogonal quality of the meshing are within the acceptable region of simulation accuracy.

TABLE I
QUALITY OF MESH FOR ALL MODELS

Model	Nodes	Elements	Skewness	Orthogonal Quality
Smooth surface	142564	782743	0.222	0.862
DR = 0.05	319366	1798765	0.221	0.863
DR = 0.2	282862	1578882	0.221	0.863
DR = 0.3	283411	1580105	0.221	0.863
DR = 0.4	284911	1586428	0.221	0.863
DR = 0.5	280090	1560387	0.221	0.863

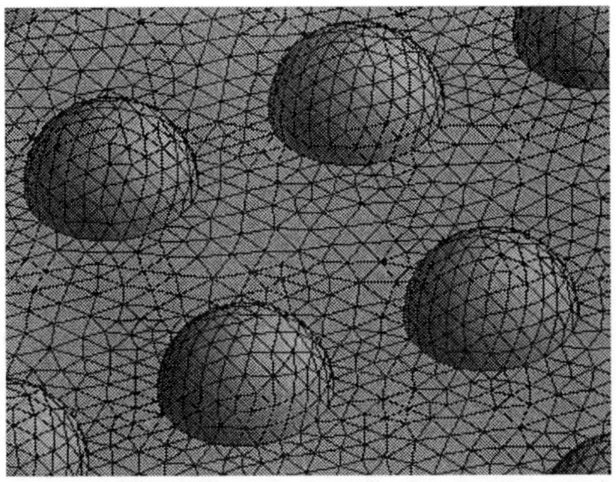

Fig. 6 Tetrahedral mesh around dimples

D. Turbulence Model

In this paper, *k-ε* model in ANSYS Fluent has been chosen for turbulence modeling. *k-ε* model consists of two transport equations to represent the turbulent properties of the flow. The first transport variable is the kinetic energy term, *k*. It is the variable to determine the turbulent energy production.

$$\partial/\partial t\,(\rho k) + \partial/\partial x_i\,(\rho k u_j) = \partial/\partial x_j\,[(\mu + \mu_t/\sigma_k).\partial/\partial x_j] + G_k + G_b - \rho\varepsilon - Y_M + S_k \quad (2)$$

The second transport variable is the turbulent dissipation term, ε. It is the variable to determine the scales of the turbulence.

$$\partial/\partial t\,(\rho\varepsilon) + \partial/\partial x_i\,(\rho\varepsilon u_j) = \partial/\partial x_j\,[(\mu + \mu_t/\sigma_\varepsilon)\,\partial_\varepsilon/\partial x_j] + \rho C_1 S_\varepsilon - \rho C_2[\varepsilon^2/(k + \sqrt{v\varepsilon})] + C_{1\varepsilon}.\varepsilon/k.C_{3\varepsilon}.G_b + S_\varepsilon \quad (3)$$

In the equations above, G_k is the generation of turbulence kinetic energy due to the mean velocity gradients. G_b is the generation of turbulence kinetic energy due to buoyancy. Y_M is the contribution of the fluctuating dilatation in compressible turbulence to the overall dissipation rate. $C_{1\varepsilon}$andC_2,and $\sigma_k$andσ_ε are constants and turbulent Prandtl numbers for *k* and ε respectively. $S_k$andS_ε are user defined source terms.

E. Boundary Conditions

In this work, the air velocity has been set to 40 m/s at the inlet that gives the Reynolds number as;

$$Re = \frac{\rho UA}{\mu} = \frac{1.2\times40\times0.065741}{1.789\times10^{-5}} = 176387 \quad (4)$$

where ρ is the density of air; is the flow velocity; *A* is the frontal area of the car model; μ is the dynamic viscosity of air.

III. Results

For the model without application of dimple, there is insignificant turbulent kinetic energy on the car surface (Fig. 7 (a)). When compared with the model without dimple, turbulent kinetic energy is generated within the dimple and at the vicinity of the dimple edge (Figs. 7 (b)-(f)). These results suggest that the boundary layer flows manage to go further before the flow separation takes place. The geometry of dimple changes the kinematics and dynamics of flow. Complex interaction between the turbulent fluctuating flow and the mean flow escalates the turbulence quantities and the boundary layer gains momentum from this turbulent kinetic production. The maximum level of turbulent kinetic energy occurs at DR = 0.4 as shown in Fig. 7 (e).

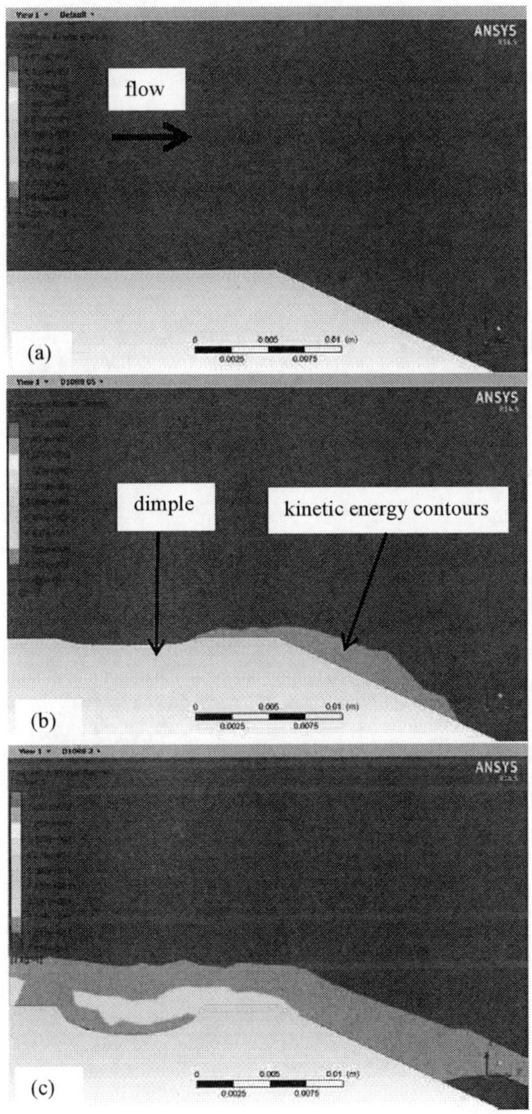

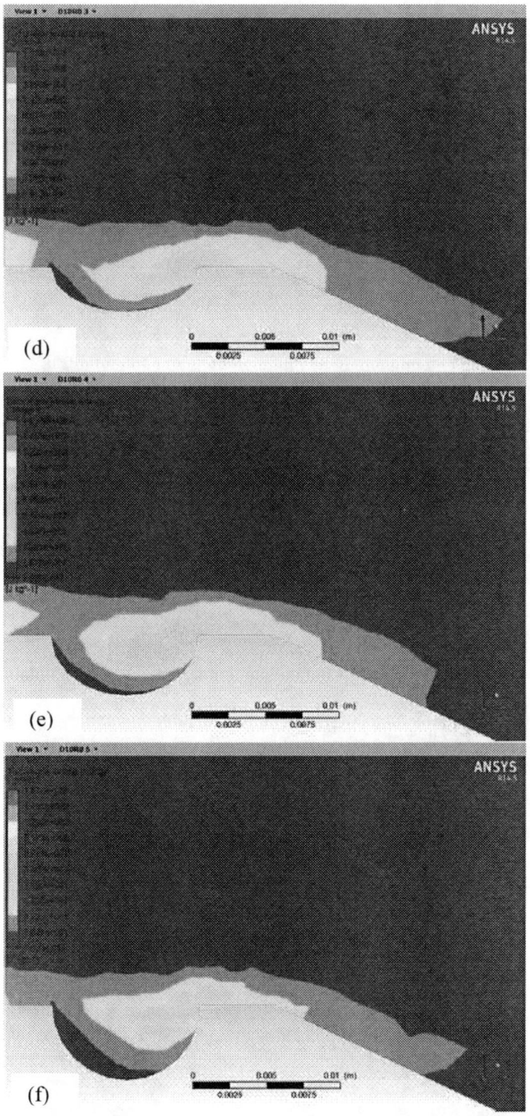

Fig. 7 Turbulence kinetic energy contours (a) DR = 0; (b) DR = 0.05; (c) DR = 0.2; (d) DR = 0.3; (e) DR = 0.4; (f) DR = 0.5

TABLE II
COMPARISON OF DRAG COEFFICIENT

Model	Drag Coefficient	Percentage Difference (%)
Smooth surface	0.261	-
DR = 0.05	0.257	1.32
DR = 0.2	0.259	0.55
DR = 0.3	0.260	0.07
DR = 0.4	0.255	1.95
DR = 0.5	0.258	0.84

IV. CONCLUSION

All the dimple ratio models give smaller coefficient of drag, C_D compared to the model without dimple (Table II). The coefficient of drag, C_D is reduced by 1.95% for the model with DR = 0.4. The results are encouraging since the simulation is only based on one dimple. Different parameters like dimples position, number of dimples, dimples orientation will be tested in order to fully understand the performance of the dimple application on vehicle aerodynamics.

REFERENCES

[1] Ahmed, S.R., Ramm, G. & Faltin G. (1984). Some salient features of the times-averaged ground vehicle wake. SAE Society of Automotive Eng., Inc, 1(840300):1 – 31.

[2] Dol, S.S., Kopp, G.A. & Martinuzzi, R.J.M. (2008). The suppression of periodic vortex shedding from arotating circular cylinder. Journal of Wind Engineering and Industrial Aerodynamics 96, 1164 – 1184.

[3] Hermann, L., Michael, B. & Cagatay, K. (2008). Drag Reduction by dimples? A Complementary Experimental/Numerical Investigation. International Journal of Heat and Fluid Flow 29(3), 783 - 791.

[4] Munson, B.R, Young, D.F. & Okiishi, T.H. (2002). Fundamentals of Fluid Mechanics, 4th Edition, John Wiley and Sons.

[5] Srivastav, D. (2012) Flow control over airfoils using different shaped dimples. International Conference on Fluid Dynamics and Thermodynamics Technologies (p. 92-97). Singapore: IACSIT Press.

[6] White, F.M. (2003). Fluid Mechanics, 5th Edition McGraw Hill.

World Academy of Science, Engineering and Technology
International Journal of Mechanical, Aerospace, Industrial, Mechatronic and Manufacturing Engineering Vol:9, No:6, 2015

Aerodynamic Prediction and Performance Analysis for Mars Science Laboratory Entry Vehicle

Tang Wei, Yang Xiaofeng, Gui Yewei, Du Yanxia

***Abstract*—**Complex lifting entry was selected for precise landing performance during the Mars Science Laboratory entry. This study aims to develop the three-dimensional numerical method for precise computation and the surface panel method for rapid engineering prediction. Detailed flow field analysis for Mars exploration mission was performed by carrying on a series of fully three-dimensional Navier-Stokes computations. The static aerodynamic performance was then discussed, including the surface pressure, lift and drag coefficient, lift-to-drag ratio with the numerical and engineering method. Computation results shown that the shock layer is thin because of lower effective specific heat ratio, and that calculated results from both methods agree well with each other, and is consistent with the reference data. Aerodynamic performance analysis shows that CG location determines trim characteristics and pitch stability, and certain radially and axially shift of the CG location can alter the capsule lifting entry performance, which is of vital significance for the aerodynamic configuration design and inner instrument layout of the Mars entry capsule.

***Keywords*—**Mars entry capsule, static aerodynamics, computational fluid dynamics, hypersonic.

I. Introduction

MOTIVATED by the curiosity about the universe origin, the search for the extraterrestrial intelligence, and the interest in the space recourses exploitation, exploration missions to Mars have been carried out several times since 1960s. Mars landing probe is surrounded with non-air gas during its entry into the Martian atmosphere. Hypersonic flow exists around the probe, and as a result severe aerodynamic force and aero-heating occur on the probe surface. Therefore, specific aerodynamic and thermal protection problems are brought forth. Mars entry missions have some similarity to Earth reentry missions, but remarkable difference exists at the same time.

The recent Mars entry mission is called Mars Science Laboratory (MSL), which successfully landed on the Mars surface on August 2012.The MSL entry capsule traveled the entire Martian atmosphere before descending and landing [1], [2], and selected complex lifting entry for sufficient deceleration and precise landing performance. The hypersonic lift-to-drag (L/D) is designed for 0.24 at the trimmed angle of attack of 16°.The capsule center of gravity (CG) is radially offset to meet the demand of the trimmed angle of attack and L/D [3]. Such entry technique was also adopted by most of the Earth reentry capsules, such as Soyuz. How to precisely predict the lifting entry process is one of the most important problems for Mars entry capsule design.

Though several Mars entry missions have been achieved, small amount of flight data is available. In addition, there exist no ground-based facilities that can reproduce the high-speed and high-temperature flow close to the Martian environment. The computational fluid dynamics technique, therefore, turns out to be a significant approach for flight aerodynamic analysis. For hypersonic Mars entry, Navier-Stokes equations solver is believed to be applicable for flowfield prediction and aerodynamic performance analysis, because the peak surface pressure occurs in the continuum regime.

Up to now, lots of numerical works for hypersonic flow have been made to evaluate the aerodynamics for different configurations. Most of the numerical works used the complicated chemical and thermal non-equilibrium model, which have not been effectively verified by experiments [4]. Therefore, some computational attempts based on the perfect gas model have been made to avoid the uncertainty of the chemical reacting model, and comparative discussions with the complex reacting model were also performed [5]-[8]. For the Mars entry process, because the atmosphere differs from the air, the real post-shock effective specific heat ratio $\gamma 2$is less than that of reentry process [5]. Therefore, the specific heat ratio is no longer assumed to be1.4 for Mars entry process, and aerodynamic coefficients may thus alter in the Mars entry case. The specific heat ratio is mainly the function of temperature, as is demonstrated in statistical thermodynamics [9]. We can use the effective specific heat ratio to simulate the hypersonic non-air flow with high temperature real-gas effects [10]. Such approach has been widely used in the rocket combustion flow, hypersonic reentry flow [11], [12] and so on.

In this paper, we developed the three-dimensional numerical method for precise computation and the surface panel method for rapid engineering prediction. A serial of numerical and engineering simulations for the MSL entry capsule were performed at different angle of attack at a given flight state based on the specified effective specific heat ratio assumption. Detailed flowfield analysis and the static aerodynamic performance were then discussed to provide the aerodynamic design suggestions.

II. Methods

A. Numerical Simulation Methods

Computational Fluid Dynamics technique was undertaken to

Tang Wei, Yang Xiaofeng, and Du Yanxia are with the State Key Laboratory of Aerodynamics, China Aerodynamics Research and Development Center, Mianyang Sichuan 621000,China (e-mail: Tangwei@cardc.cn, yangxiaofeng@cardc.cn, yanxiadu@cardc.cn).

Gui Yewei is with the State Key Laboratory of Aerodynamics, China Aerodynamics Research and Development Center, Mianyang Sichuan 621000,China (corresponding author, e-mail: guiyewei777@126.com).

predict the aerodynamic force for the Mars entry vehicle. A finite volume approach was used to solve the full Navier–Stokes flowfield equations for a calorically perfect gas. The code used the van Leer flux-vector splitting method for the inviscid fluxes with the 2nd order correction using the van Leer limiter. For time integration, the non-iterative implicit method was used for rapid convergence.

Constant inflow condition was imposed on the far field boundary, and the extrapolation was used on the outflow boundary in all solutions. Non-slip wall boundary condition was implemented and the thermal state of the surface is radiative equilibrium with a fixed surface emissivity of 0.78.

The flowfield is assumed to be steady and laminar. Unsteady effects are weak at these high speeds for long duration entry flight [13], and moreover, unsteady aerodynamic contribution is negligibly small in the afterbody [14]. Transition to turbulence could occur but is weak due to low Reynolds number in the relatively thin Martian atmosphere [4].

High-temperature gas viscosity coefficient is assumed to depend only on temperature, and can be specified by Sutherland's law [15]. The dynamic viscosity coefficient μ for Martian gas is:

$$\frac{\mu}{\mu_0} = \left(\frac{T}{T_0}\right)^{\frac{3}{2}} (T_0 + C)/(T + C) \quad (1)$$

where the reference viscosity μ_0 = 1.48×10^{-5} kg/(m·s), the reference temperature T_0 = 293.15 K, and the Sutherland constant C = 240 K. Sutherland's law agrees well with theoretical calculation from Fenghour [16].

For high-temperature Martian gas, we assume Pr = 0.71, which is used to obtain the thermal conductivity $k=\mu c_p/Pr$.

B. Engineering Prediction Method

Surface panel method was used to calculate the surface pressure coefficient for rapid prediction. The pressure coefficient for each individual element is the function of inflow velocity vector, the element area and the surface normal vector of the given element, and has no relation with the neighbor element. The total aerodynamic coefficients are then calculated by integrating the panel pressure coefficient along the total surface.

The main work of the surface panel method is to calculate the surface pressure coefficient. Multiple engineering methods to predict the surface pressure coefficient have been developed so far, with their peculiar applications. The modified Newtonian theory is believed to be quite suitable for hypersonic aerodynamics prediction of blunt body, such as reentry shapes, Mars entry capsules, etc. The modified Newtonian formula can be described as:

$$C_p = C_{pmax}(\bar{V}_\perp / V_\infty)^2 \quad (2)$$

where C_{pmax} is the stagnation pressure coefficient, described as:

$$C_{pmax} = \frac{2}{\gamma Ma_\infty^2}\left[\left(\frac{(\gamma+1)^2 Ma_\infty^2}{4\gamma Ma_\infty^2 - 2(\gamma-1)}\right)^{\frac{\gamma}{\gamma-1}} \cdot \left(\frac{1-\gamma+2\gamma Ma_\infty^2}{\gamma+1}\right) - 1\right] \quad (3)$$

It can be seen that the stagnation pressure coefficient is the only function of inflow Mach number and the specific heat ratio.

It is noted that the modified Newtonian formula is valid for windward region, and the coefficient in the leeward region is assumed to be zero because the pressure effect here is negligibly weak. The velocity ratio is the cosine of the angle between the inflow velocity and the surface normal vector.

C. Effective Specific Heat Ratio Method

Multicomponent mixture exists in the Martian atmosphere, including approximately95% CO2, 3% N2, 2% Ar and other negligible species. Since the real values of specific heat ratio markedly decrease at high-Mach-number flight conditions when the gas goes across the shock, effective specific heat ratio was specified to take the non-air and high-temperature real-gas effects into account. The post-shock specific heat ratio, which only depends on the local temperature for thermal perfect gas, is expected to be the effective value γeff. Considering the vibrational excitation, the local specific heat ratio is:

$$\gamma = \sum_{j=1}^{m} w_j c_{P,j} \,/\, \sum_{j=1}^{m} w_j c_{V,j} \quad (4)$$

where m is the number of gas species, w_j is the mass fraction of the species j. For each species, c_p and c_v are determined by molecular transitional, rotational and vibrational energy, so we have

$$c_V = \left(\frac{1}{2}n_t + \frac{1}{2}n_r + \sum_{i=1}^{n_v} \frac{\left(\frac{T_{v,i}}{T}\right)^2 e^{\frac{T_{v,i}}{T}}}{\left(e^{\frac{T_{v,i}}{T}}-1\right)^2}\right) R \quad (5)$$

$$c_P = c_V + R$$

where n_t, n_r and n_v, are the molecular transitional, rotational and vibrational degrees of freedom, and $T_{v,i}= hv_i/k$ is the vibrational characteristic temperature, determined by the vibrational characteristic frequency v_i referred by [**].

The local post-shock temperature can be simply obtained by solving the generalized Rankine-Hugoniots hock relation. For normal shock, we have:

$$\left(\frac{\rho_2}{\rho_\infty}\right)^2 \left[M_\infty^2 + \frac{2}{\gamma_\infty - 1}\right] + \left(\frac{\rho_2}{\rho_\infty}\right)\left[-\frac{2\gamma_2(\gamma_\infty M_\infty^2+1)}{\gamma_\infty(\gamma_2-1)}\right] + \left[M_\infty^2 \frac{\gamma_2+1}{\gamma_2-1}\right] = 0 \quad (6)$$

$$\frac{\frac{p_2}{p_\infty}-1}{\gamma_\infty M_\infty^2} = \frac{\frac{p_2}{p_\infty}-1}{\frac{p_2}{p_\infty}} \quad (7)$$

$$\frac{T_2}{T_\infty} = \frac{p_2}{p_\infty}\frac{\rho_\infty}{\rho_2}\frac{R_\infty}{R_2} \quad (8)$$

By iterating formulae (4) to (8), we obtain the local specific heat ratioγ_2 in the shock layer, which is considered to be the effective specific heat ratio γ_{eff}. It must be noted that the effective value is valid only in the stagnation region, but we can generalize it to the overall flowfield because the stagnation aeroforce is dominant and such difference is negligibly small. The effective specific heat ratio contours with the inflow Mach number and temperature are shown in Fig. 1. In the infinite case

($M\infty \rightarrow \infty$), γ_{eff} tends to a specific value because of full excitation of vibrational energy and the stagnation pressure coefficient gets independent on the inflow Mach number.

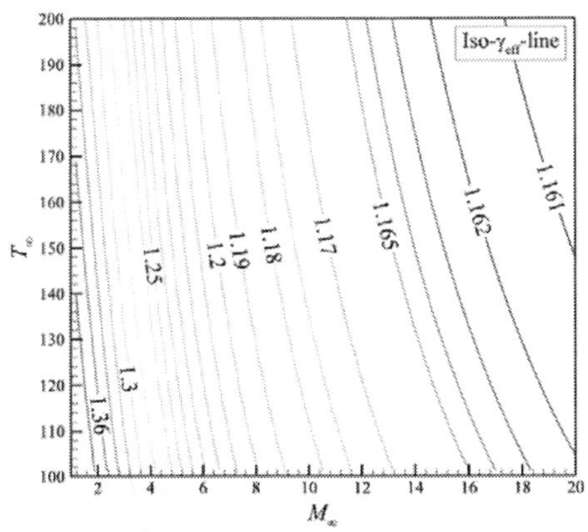

Fig. 1 Effective specific heat ratio contours with inflow Mach number and temperature

D. Validation Case

The validation model is a scaled 70° sphere-cone configuration with the available numerical and experimental data from [17], [18]. The wind tunnel experiment use CO_2 as the test gas with the inflow enthalpy of 1.89MJ/kg (U_∞ = 1908 m/s, p_∞ = 1010 Pa). Calculated results from numerical simulation and engineering prediction based on the effective specific heat ratio method are compared against the experimental data. The calculation results of the surface pressure from both two mentioned approaches, the DPLR data and the experimental data are all shown in Fig. 2 for comparison. The comparison highlights a good agreement between calculated and experimental data, thus confirming reliability of the numerical simulations and engineering prediction for Mars entry capsules.

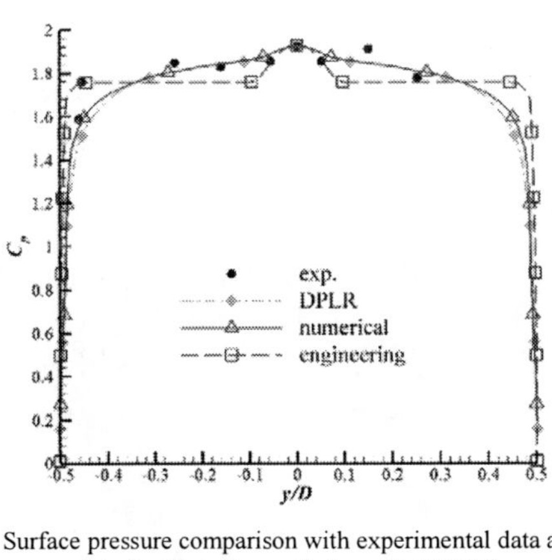

Fig. 2 Surface pressure comparison with experimental data and DPLR data for the inflow enthalpy of 1.89 MJ/kg

III. Results and Discussion

A. Geometry and Grid Generation

The proposed geometry for aerodynamic performance analysis is the Mars Science Laboratory entry vehicle with a 70-degree sphere-cone forebody and three-sectional-cone afterbody as described in Fig. 3. The body axis system is used for aerodynamic performance analysis also shown in Fig. 3, in which system a pitch-up motion means a positive angle of attack (α> 0), and a positive pitch moment pitches the capsule up.

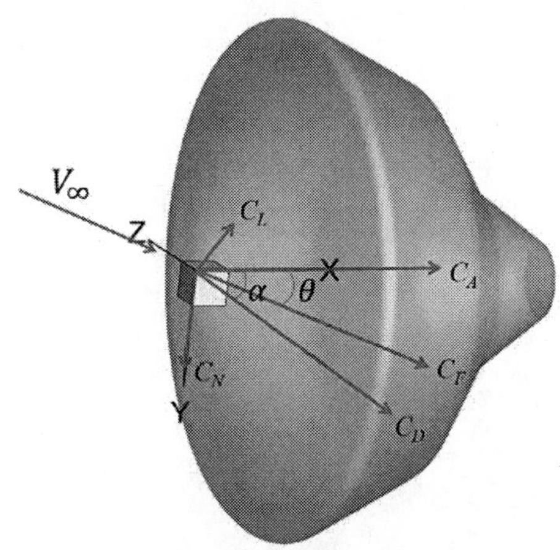

Fig. 3 MSL geometry and definition of aerodynamic forces

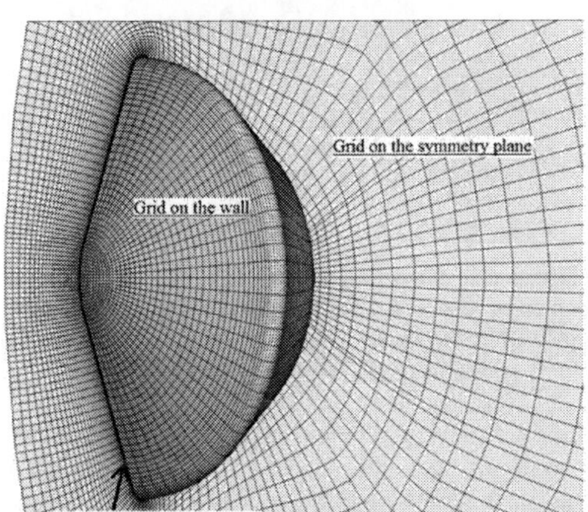

Fig. 4 MSL computational grid schematic (the actual grid is twofold refined in each direction)

Two-dimensional surface grid is required for rapid aerodynamic prediction using the panel method, while full three-dimensional grid is necessary for numerical simulation of overall flowfield. The computational grid for MSL configuration is a multi-block structured grid, mainly growing from the surface grid taking the boundary layer into consideration, and has an overall number of approximately 1.4

million cells for half body. To assure accurate prediction, the grid is everywhere orthogonal to the body at the surface. Grid independence analysis shows that grids for the wall cell Reynolds number $Re_c=\rho_\infty U_\infty d_w/\mu_\infty$ can fulfill the accuracy requirement of aerodynamics analysis, where d_w is the normal grid distance close to the wall (the wall step). The wall cell Reynolds number for the current analysis is hence 6.4. Fig. 4 shows both the two-dimensional surface grid and the grid on the symmetry plane.

B. Mars Entry Flowfield

In order to figure out the detailed hypersonic flowfield, steady numerical simulations for a specific hypersonic flight state are herein performed. The flight speed is 5164 m/s and the Mach number is 26.1 with an angle of attack $\alpha = -17.05°$, so that the inflow unit Reynolds number $Re_\infty=2.4\times10^6$/m. The effective value of specific heat ratio used in the calorically perfect gas flow simulation is derived to be 1.16.

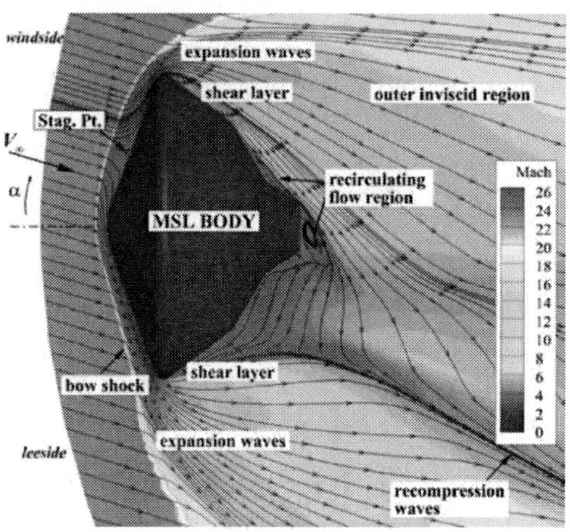

Fig. 5 Hypersonic flow structure for MSL on the symmetry plane

Hypersonic flowfield for the MSL entry vehicle is complicated, especially for the vortex flow. Fig.5 shows the main steady structure of the hypersonic flow on the symmetry plane. A strong bow shock wave is detached from the surface and lies very close to the heat shield. The gas across the shock wave is compressed to form a shock layer, much thinner than that of the air flow. Inflow uniform streamlines are deflected across the shock and travel around the body (Fig. 6). Expansion waves occur near the shoulder due to large deflection angle. The gas gets accelerated in this region and merges into the external inviscid flow downstream from the shoulder. The heat shield boundary layer thickness increases from the stagnation point to the shoulder, and the shear layer appears after the shoulder, resulting in local vortex motion. Consequently, flow separation arises on the backshell (Fig. 6). A large region of recirculating flow leads to relatively low pressure distribution on the backshell, which has a negligible contribution to the total coefficients. The wake flow after the capsule meets and then accumulates to form the recompression waves, which are nearly parallel to the inflow direction, and grow weaker to the outer flow.

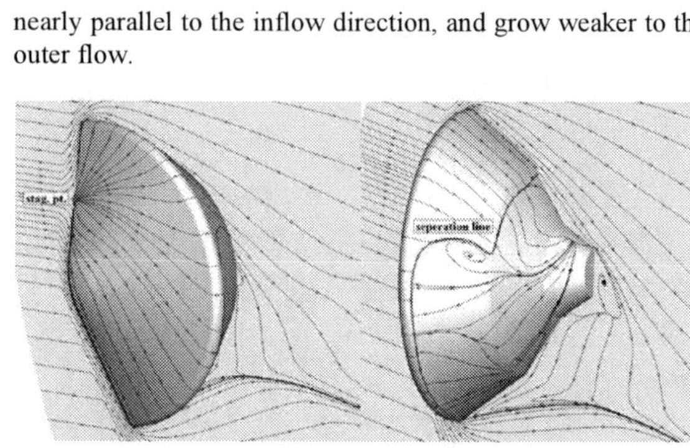

Fig. 6 Streamlines on the symmetry plane and near the surface of MSL

C. Surface Pressure Distribution

For the same inflow boundary condition, rapid engineering calculation was performed to obtain the surface pressure and the total aerodynamic coefficients. The calculated surface pressure coefficient distribution is shown in Fig. 7. Compared with the numerical results, the rapid prediction results regularly spread on the cone frustum from the heat shield apex, due to the single dependency of pressure coefficient on geometry angle.

Fig. 8 shows the heatshield surface pressure coefficient distribution on the symmetric line, including the numerical results, engineering prediction values and the reference data [3] for comparison. As is expected, the surface pressure is the highest at the stagnation point and falls off rapidly as the flow expands around the shoulder and onto the afterbody. On the whole, calculated results for surface pressure from numerical simulation and engineering calculation keep in good agreement with the reference, especially on the windward region, which dominates the total aeroforce coefficients.

It is noted that a certain degree of fluctuation exists for numerically calculated surface pressure in the heatshield leeside, partly because of extremely high Mach number and very low specific heat ratio. Nevertheless, the overall influence of such fluctuation is relatively weak after integration for aerodynamic analysis. It maintains the level of surface pressure on the cone frustum because the geometric angle between the wall and inflow velocity holds on the frustum.

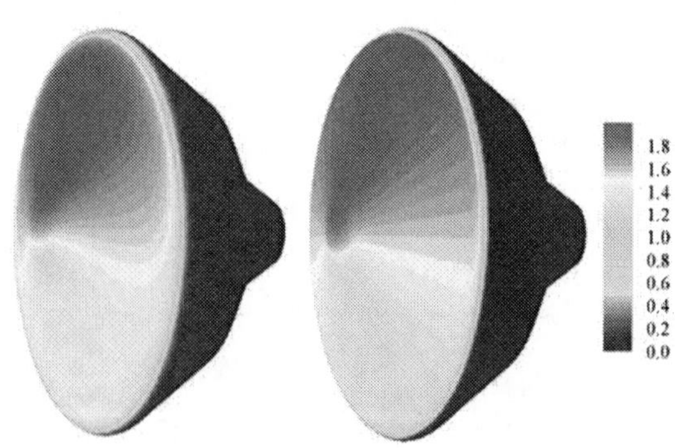

(a) numerical (b) engineering

Fig. 7 Surface pressure distribution

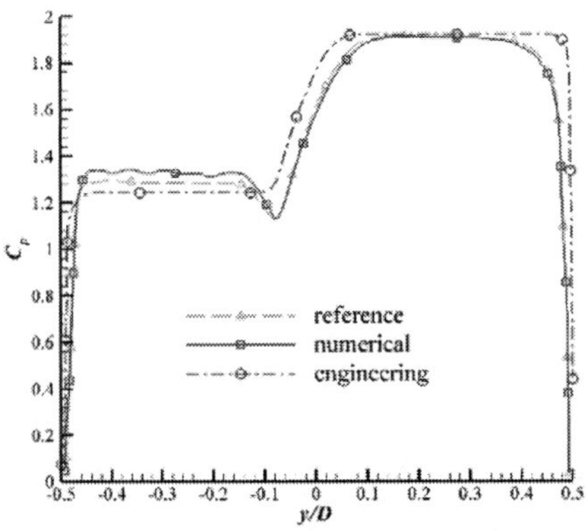

Fig. 8 Surface pressure distribution on symmetry plane of MSL heat shield

The stagnation point locations, and the shock and sonic lines on the symmetric plane are also shown in Fig. 9. For angle of attack of -6°, the stagnation point is located near the axis on the sphere cone; when the angle of attack increases to -17.5°, the stagnation point is moved far away from the axis to the cone frustum section; when to -29°, the stagnation point is moved very near to the shoulder. The windward sonic line merges into the boundary layer at the shoulder, while the leeward one at different place depending on the angle of attack. For lower angle of attack, -6° for example, the sonic line merges into the boundary layer at the shoulder, the same as the windward one. When the angle of attack increases, the merging location moves to the sphere-cone junction (in -17.5° case), even near the axis (in -29° case), and the sonic line has an inflection in the vicinity of the apex because of strong compressibility. The high pressure region is surrounded by the windward sonic line and the wall, which dominates the total aerodynamic coefficients, particularly at small angles of attack.

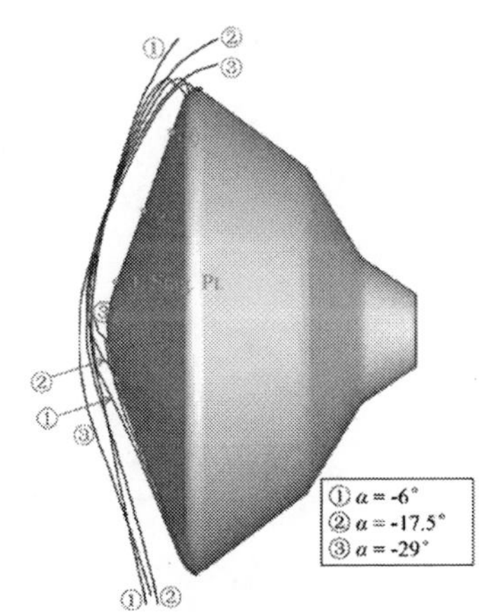

Fig. 9 Shock wave and sonic line for different angle of attack

D. Aerodynamic Coefficients

The integrated lift (C_L) and drag (C_D) coefficients and the lift-to-drag ratio (*L*/*D*), are shown in Fig. 10 as a function of angle of attack from zero to -45° for both numerical and engineering results. Both results keep the same trend as to the angle of attack, and the values of them are quite close in low angle of attack. At large angle of attack, the windward flow differs remarkably from the leeward flow, which cannot be captured by surface panel method, so that the total coefficients have much different. C_D is predicted to decrease continuously as the magnitude of the angle of attack increases and C_L is just the reverse in the relatively low angle of attack. As a result, *L*/*D* increases with the increase of the magnitude of the angle of attack. It can be expected from computations that a trimmed *L*/*D* of 0.27 can be achieved if a trimmed angle of attack of 17.5°holds. Moreover, C_D is expected to be dominant in all total coefficients, which is the typical character of blunt entry capsule.

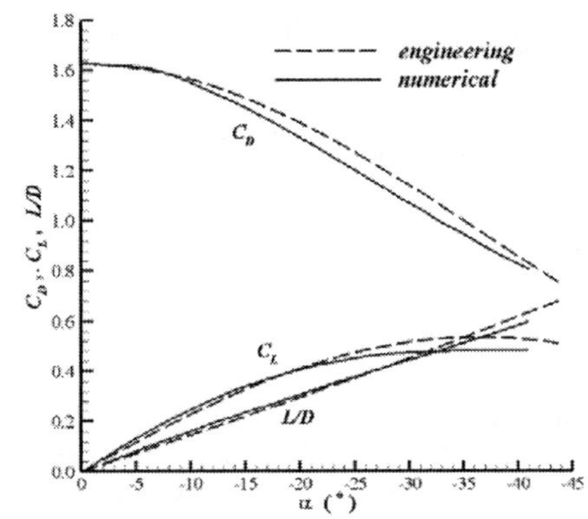

Fig. 10 The total coefficient curves for both numerical and engineering results

E. Center of Gravity and Stability Analysis

For hypersonic lifting entry process, there exists a close relationship between the center of gravity (CG) location and the flight performance with can be described by:

$$X_{cg} = X_{cp} - \Delta X\,,\quad Y_{cg} = Y_{cp} - \Delta Y\,,$$

$$\Delta X = \frac{\frac{dC_M}{d\alpha} + C_A\frac{dY_{cp}}{d\alpha} - C_N\frac{dX_{cp}}{d\alpha}}{\frac{dC_N}{d\alpha} - \frac{dC_A}{d\alpha}\frac{C_N}{C_A}}\,,\quad \Delta Y = \frac{C_N}{C_A}\Delta X \tag{9}$$

The formula sketches the relationship between the CG location (the offset component Y_{cg} and the shift component X_{cg}) and flight requirements, including the trim characteristics (the trimmed α and *L*/*D*) and the static pitch stability (the derivative $C_{M,\alpha}$). See [19] for detailed deriving process.

Fig. 11 shows the trimmed α or *L*/*D* and the static pitch stability margin distribution in the CG plane. If we have determined the requirements of the trim condition and the static pitch stability, the needed CG position is then exactly located from Fig. 11, which help design the CG of the capsule in the aspect of aerodynamics. Considering the MSL entry capsule, CG is located near the volumetric center (30~50% of the total length) if the capsule mass is relatively uniformly distributed. It can be seen from the figure that the capsule is statically pitch stable for the small trimmed *L*/*D*, and the stability margin is about -0.003/°. The capsule is radially offset for a given degree in the demand of the hypersonic trim condition.

In addition, a small radial bias of CG may cause remarkable flight performance deviation for lifting entry, which should be paid close attention to in the process of aerodynamic design.

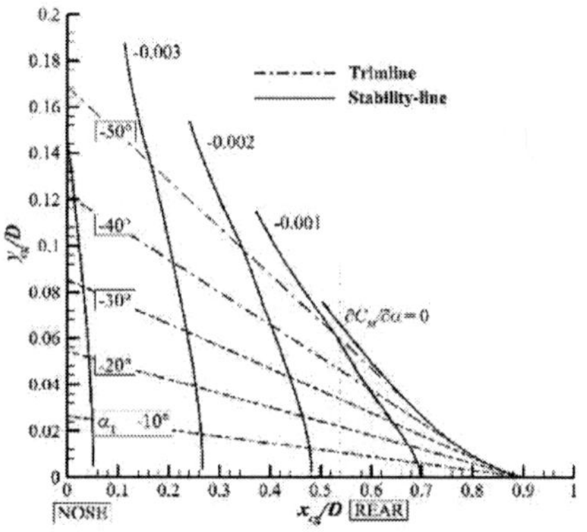

Fig. 11 The trimmed α or L/D and the static pitch stability margin distribution in the CG plane (solid: stability margin isoline, dashed-dot: trimmed α isoline, with the radial direction zoomed in)

IV. Conclusions

This paper paid great attention to the numerical method for precise simulation of hypersonic Martian flowfield and surface panel method for rapid prediction of aerodynamic coefficients. Detailed flowfield analysis for Mars exploration mission was performed by carrying on a series of fully three-dimensional Navier–Stokes computations. The static aerodynamic performance was then discussed, including the surface pressure, lift and drag coefficient, lift-to-drag ratio with the numerical and engineering method. Calculated results from both methods agree well with each other, and are consistent with the reference data.

Aerodynamic performance analysis shows CG position determines trim characteristics and stability, which can be guidance on aerodynamic design. Specific CG position can be selected so as to fulfill the hypersonic lifting entry. Consequently, proper aerodynamic configuration and cabin equipment layout is needed for CG adjustment to meet the static aerodynamics requirements. This investigation provides researchers with guidelines for the optimization of the design of the Mars entry capsule.

Acknowledgments

This research was supported by National Natural Science Foundation of China (No. 91216204, 11472295, 11302246).

References

[1] R.D. Braun, R.M. Manning, Mars exploration entry, descent, and landing challenges, J. Spacecr. Rockets 44 (2) (2007) 310-323.

[2] R. Prakash, P.D. Burkhart, A. Chen, K.A. Comeaux, C.S. Guernsey, D.M. Kipp, D.W. Way, Mars Science Laboratory entry, descent, and landing system overview, in: Proceedings of Aerospace Conference, IEEE, 2008, pp. 1-18.

[3] K.T. Edquist, A.A. Dyakonov, M.J. Wright, C.Y. Tang, Aerothermodynamic environments definition for the Mars Science Laboratory entry capsule, AIAA Pap. 1206 (2007) 8-11.

[4] M.J. Wright, C.Y. Tang, K.T. Edquist, B.R. Hollis, P. Krasa, C.A. Campbell, A review of aerothermal modeling for Mars entry missions, AIAA Pap. 443 (2010) 4-7.

[5] P.A. Gnoffo, K.J. Weilmuenster, R.D. Braun, C.I. Cruz, Effects of sonic line transition on aerothermodynamics of the Mars pathfinder probe, in: Proceedings of AIAA, 1995, 1825-CP.

[6] D.K. Prabhu, D.A. Saunders, On heatshield shapes for Mars entry capsules, in: Proceedings of AIAA, 2002, 1221.

[7] A. Viviani, G. Pezzella, Aerodynamic analysis of a capsule vehicle for a manned exploration mission to Mars, in: Proceedings of the 16th AIAA/DLR/DGLR International Space Planes and Hypersonic Systems and Technologies Conference, AIAA, 2009, 7386.

[8] P.A. Liever, S.D. Habchi, S.I. Burnell, J.S. Lingard, CFD prediction of the Beagle 2 aerodynamic database, in: AIAA (2002) 0683.

[9] E.H. Hirschel, C. Weiland, Selected aerothermodynamic design problems of hypersonic flight vehicles, Springer Press, New York, 2009.

[10] J.D. Anderson, Jr., Hypersonic and high-temperature gas dynamics, AIAA, 2006.

[11] G.J. Brauckmann, J.W. Paulson Jr., K.J. Weilmuenster, Experimental and computational analysis of shuttle orbiter hypersonic trim anomaly, J. Spacecr. Rockets 32 (5) (1995) 758–764.

[12] K.J. Weilmuenster, H.H. Hamilton, A comparison of computed space shuttle orbiter surface pressures with flight measurements, AIAA, 1982, 0937.

[13] R.A. Mitcheltree, P.A. Gnoffo, Wake flow about the Mars pathfinder entry vehicle, J. Spacecr. Rockets 32.5 (1995) 771-776.

[14] K.T. Edquist, P.N. Desai, M. Schoenenberger, Aerodynamics for the Mars Phoenix entry capsule, in: Proceedings of AIAA/AAS Astrodynamics Specialist Conference and Exhibit, AIAA, 2008, 7219.

[15] R.C. Weast, CRC Handbook of Chemistry and Physics, 65th ed. Chemical Rubber Company Press, Cleveland, 1984.

[16] A Fenghour, W.A. Wakeham, V. Vesovic, The viscosity of carbon dioxide, J. Phys. Chem. Ref. Data 27 (1) (1998) 31–44.

[17] M.S. Holden, T.P. Wadhams, G.J. Smolinski, M.G. MacLean, J. Harvey, B.J. Walker, Experimental and numerical studies on hypersonic vehicle performance in the LENS shock and expansion tunnels, AIAA Pap. 125 (2006) 2006.

[18] M. MacLean, M.S. Holden, Catalytic effects on heat transfer measurements for aerothermal studies with CO2, AIAA Pap. 182 (2006) 9-12.

[19] X. Yang, W. Tang, Y. Gui, Y. Du, G. Xiao, L. Liu,Hypersonic static aerodynamics for Mars science laboratory entry capsule, Acta Astronaut. 103 (2014) 168-175.

Half Model Testing for Canard of a Hybrid Buoyant Aircraft

A. U. Haque, W. Asrar, A. A. Omar, E. Sulaeman, J. S. Mohamed Ali

***Abstract*—Due to the interference effects, the intrinsic aerodynamic parameters obtained from the individual component testing are always fundamentally different than those obtained for complete model testing. Consideration and limitation for such testing need to be taken into account in any design work related to the component buildup method. In this paper, the scaled model of a straight rectangular canard of a hybrid buoyant aircraft is tested at 50 m/s in IIUM-LSWT (Low Speed Wind Tunnel). Model and its attachment with the balance are kept rigid to have results free from the aeroelastic distortion. Based on the velocity profile of the test section's floor; the height of the model is kept equal to the corresponding boundary layer displacement. Balance measurements provide valuable but limited information of overall aerodynamic behavior of the model. Zero lift coefficient is obtained at -2.2° and the corresponding drag coefficient was found to be less than that at zero angle of attack. As a part of the validation of low fidelity tool, plot of lift coefficient plot was verified by the experimental data and except the value of zero lift coefficients, the overall trend has under predicted the lift coefficient. Based on this comparative study, a correction factor of 1.36 is proposed for lift curve slope obtained from the panel method.

***Keywords*—Wind tunnel testing, boundary layer displacement, lift curve slope, canard, aerodynamics.

I. Introduction

ONE way of determining the accuracy of any analytical technique is to compare data so obtained with that from a scaled down model testing in wind tunnel at comparable Reynolds number [1]. To achieve the same Reynolds number as of the real application, the kinematic viscosity or velocity has to be altered. In most wind-tunnels air at atmospheric pressure is used, and the only option left is to increase the velocity. Often it is not possible to increase much the velocity, so the results from wind-tunnel experiments are not available at the required full scaled Reynolds number. Moreover, in wind tunnel testing, the drawback is that small scale models are often not full scale ditto. This is due to the scaling issues as some minute geometric parameters are washed out [2]. One of the prospective solutions of above mentioned issues is to go for half model testing (HMT). However, HMT is also not free from the problems. The merits and de-merits of both testing methods are briefly shown in Table I.

TABLE I
COMPARISON OF HALF AND FULL MODEL TESTING

S. No.	Parameters	Half Model	Full Model
a	Operating envelop of Reynolds number	Increases	Limited due to size of geometry
b	Capturing the effect of change in wing shape	Aero data will improve for twist; either aerodynamic or geometric	Data quality is less in comparison with half model
c	Model's structural rigidity/strength or stiffness	Improves	Limited due to allowable space of test section
d	Sting-support interference effects	Absent	Have a major effect
e	Minor changes in model build	Many uncertainties will arise	Restricted uncertainties
f	Sidewall effects	Wall's boundary layer and high alpha issues	Not have high impact on results
g	Penalty in drag	Increment in drag due to undesirable flow at junction	Incorporation of drag due to additional components
j	Manufacturing cost	Less	Quite high

If we attach the wing's model with the external strut then there will be additional interferences due to strut. Also, the presence of strut does affect the flow, especially in the area which is located at its attachment. In the present study, a scaled model of canard, consisting of NACA 65(3)-218 airfoil of a hybrid buoyant aircraft is tested in IIUM-LSWT. Its geometric parameters are selected such that the exposed span and its corresponding chord are modeled by ignoring the side wall effects of the tunnel floor. Therefore, the results so obtained from the wind tunnel should not be multiplied by two as it is there for half model testing. Brief details of the configuration are given in Section II for quick reference.

XFLR, an open source code, especially for wing design has been used in the past for a good comparative study of the analytical results with those obtained from the experimental testing of a model manufactured by using the rapid prototype [16]. A good comparison of lift curve slope is obtained in this study. Authors are of the point of view that although the model was geometrically similar but it can have different aeroelastic deformations for a given combination of Reynolds number. A good numerical simulation can rule out the effect of such deformations. Hence the evidence of the success of such comparison would be useful in assessing the data quality of the model manufactured by using rapid prototyping technique. It was perhaps the motivation to use a rigid model,

A. U. Haque is with the Experimental & Computational Thermofluid Mechanics Research Group (ECTMRG), Department of Mechanical Engineering. International Islamic University Malaysia (IIUM), 50728 Kuala Lumpur, Malaysia (phone: 0060-112-6464357; email: anwar.haque@live.iium.edu.my).

W. Asrar, E. Sulaeman and J. S. Mohamed Ali are with the Experimental & Computational Thermofluid Mechanics Research Group (ECTMRG), Department of Mechanical Engineering. International Islamic University Malaysia (IIUM), 50728 Kuala Lumpur, Malaysia (e-mail: waqar@iium.edu.my, esulaeman@iium.edu.my, jaffar@iium.edu.my).

A. A. Omar is Professor, Department of Aeronautical Engineering, University of Tripoli, Libya (e-mail: aao@aerodept.edu.ly) which is tightly connected with the base plate at the tunnel floor.

II. Configuration Details

The pictorial view of the configuration is shown in Fig. 1. It contains two vertical stabilizers mounted on either side of horizontal tail, which will give better ground maneuverability of aircraft on runway, especially during taxi and take-off. Moreover, one of the unique features of the configuration is that the horizontal surface is more efficiently loaded because the vertical tails act as winglets. Horizontal tail is at some incidence angle so that positive pitching moment at zero angle of attack can be generated. This incidence decreases the overall lift which is compensated by the lift produced by the canard. The configuration contains wings similar to sailplane's wings; resulting in less induced drag from the wings and hence high glide ratio can be achieved [3].

In HB vehicle, partial lift requirement is fulfilled by the aerostatic force. Since the wings are of the reduced span and have less area owing to body lift; therefore it will result into reduced drag and weight. Additional aerodynamic lift is generated by the hull from the beginning of ground roll in take-off segment [4]. This feature of hybrid lifting hull is beneficial to meet the additional lift requirement in takeoff and perhaps will also save the additional weight requirement due to flaps.

Fig. 1 Pictorial view of hybrid buoyant aircraft

III. Geometry

The dimensions of the scaled down model of canard along with the dimensions of the test section are shown in Fig. 2. The model's span is 1 m and its chord is 0.25 m. In this way the scaled down factor is just 7.68. The exposed span of the hull is 7.68 m, which corresponds to the aspect ratio of the exposed surface that is equal to 4. The airfoil used for defining the aerodynamic contour of canard surface is NACA 65(3)-218 [5].

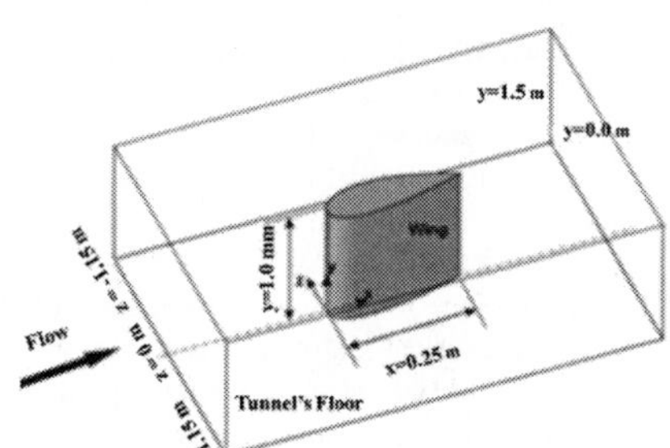

Fig. 2 Geometric details of the scaled down model in test section

IV. Experimental Setup

Static tests have been conducted in the IIUM wind-tunnel to obtain the aerodynamic response of the canard. Complete three dimensional model is mounted using a "plate mount" system which permits attachment of the model with the balance and is shown in Fig. 3. By installing a level plate, it is possible to test the model of larger scale, for which higher Reynolds numbers can be achieved. The actual Reynold's number is 3.64×10^6 (based on the mean geometric chord of the canard) and its corresponding value for sub scaled model is 8.54×10^5.

Techniques for full/half wing model for sidewall effects seem to be lacking. All that is done so far in order to minimize such effects as much as possible, by displacing the model base from the sidewall by a small distance [6]. This provides some allowance for the inward displacement of the model. However, this change does not ensure zero interference due to the inherent interaction between the model and the wall boundary layer. A number of experimental and numerical studies [7]-[13] are conducted to understand the flow topology, especially the formation of the horseshoe vortex system. However, there is no universal school of thought to define either the gap or the boundary layer suction techniques. Conducting balance measurements on wings require a minimum gap between the wing and tunnel floor to prevent the transfer of unwanted loads to the balance. Kuppa & Marchman [14] observed that a gap as small as 0.1 mm can still influence half wing measurements. Pope [10] also suggests that a small gap of the order of 0.5 % of the span may be allowed which could be a few millimeters wide for a wing with a span of the order of 1 m. Therefore, as per Pope's recommendations [10]; for one meter wing span of model of canard, the gap between the model and tunnel's floor should be equal to 5 mm. However, to the best of authors' knowledge there is no experimental evidence to support this fact. Since most of the references of experimental studies have defined some gap, therefore the absence of a gap may prevent any meaningful balance data being collected. Hence it was decided to first conduct experiments with a gap equal to the boundary layer displacement at 50 m/s.

Fig. 3 Half model tested in IIUM-LSWT

In the present work; boundary layer displacement is estimated by using the velocity profile, which was measured using boundary layer rake for 50 m/s. The rake consists of one static pressure port at the top and twenty total pressure probes, so that the dynamic pressure profiles which determine velocity profiles can be derived. Velocity profiles are then normalized with the boundary layer rake tip velocity, to force and the same is shown below as Fig. 4.

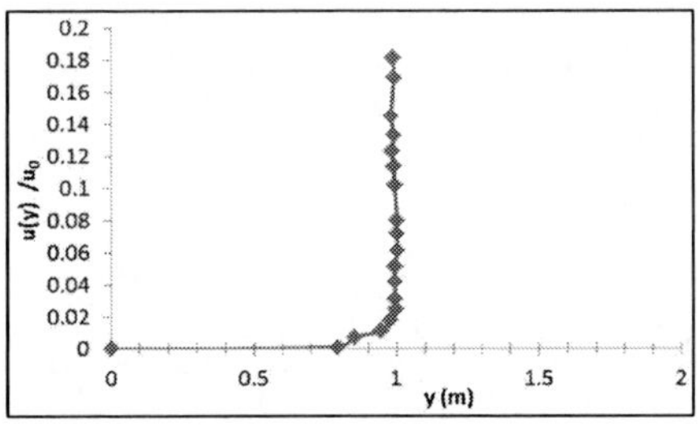

Fig. 4 Variation in ${}^{u(y)}/_{u_o}$ with respect to $y(x)$

In this figure the x-axis is the distance of the boundary layer point from the wall (*y*) and vertical axis is normalized by the corresponding measured value of the axial velocity; normalized by the maximum velocity i.e. 50 m/s. This dimensionless velocity profile shows that the velocity gradient is steeper in higher velocity regime. Boundary layer displacement δ^* is then calculated by using (1), taken from [15] and after integration, its value comes out to be equal to 1.5 mm. No adjustment for the boundary layer thickness i.e. 0.99U was carried out as actual velocity profile is now available.

$$\delta^* = \int_0^{\infty} \left(1 - \frac{u(y)}{u_0}\right) dy \qquad (1)$$

The six axis balance is used to measure the aerodynamic forces for defined range of the incidence angle i.e. $\pm$ 20^0 with increment of only one degree to get a smooth slope of the curve. Trends of variation of C_D and C_L w.r.t angle of attack (α) are shown below in Figs. 5 and 6 respectively:

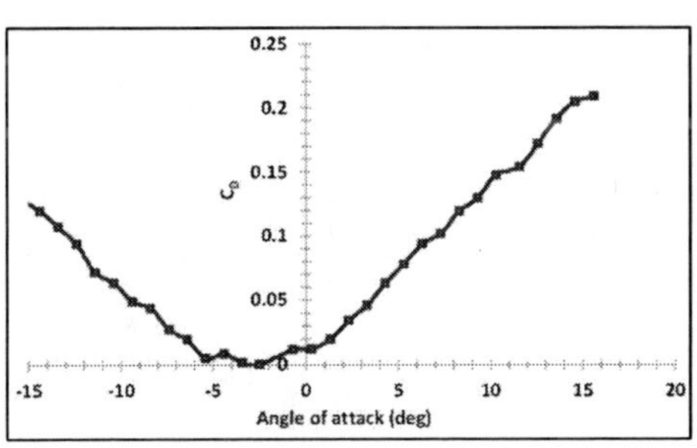

Fig. 5 Variation in C_D with respect to α

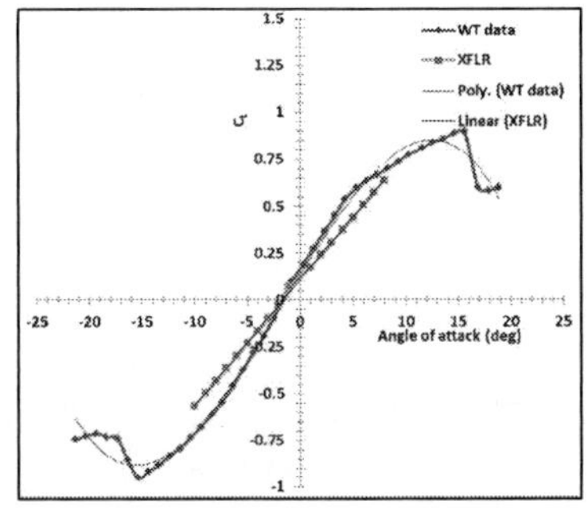

Fig. 6 Variation in C_L with respect to α

In order to compare the experimental results with the panel method; XFLR software [16] is selected. For the geometry under discussion; coordinates of NACA 65(3)-21 airfoil are first defined and batch analysis is carried out for a long range of Reynolds number in XFLR. Geometry of the wing is then defined for the required chord and span of the wing, similar to that of the sub scale model tested in the wind tunnel. VLM method is selected and lift polar is then obtained at desired velocity of 50 m/s. This method is based on the hypothesis that a lifting wing can be replaced:

"By a lifting line and that the incremental vortices shed along the span trail behind the wing in straight lines in the direction of the freestream velocity. The strength of these trailing vortices is proportional to the rate of change of the lift along the span". [17].

The results so obtained for the C_L for the defined range of α are then overlapped with the experimental results. Streamline plot and the coefficient of pressure are also plotted and the same is shown as Figs. 7 (a) and (b) respectively.

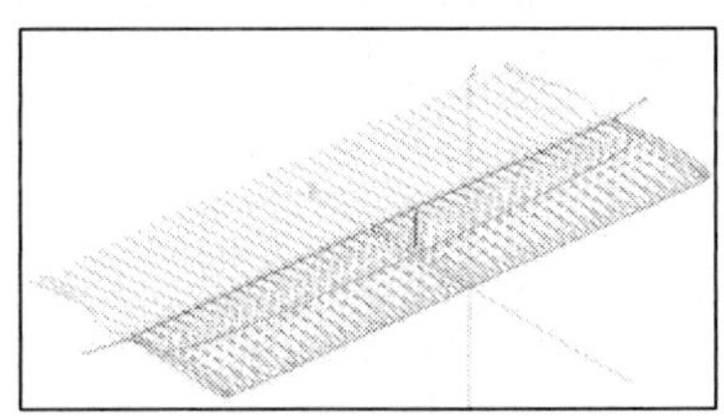

(a) Plot of streamlines

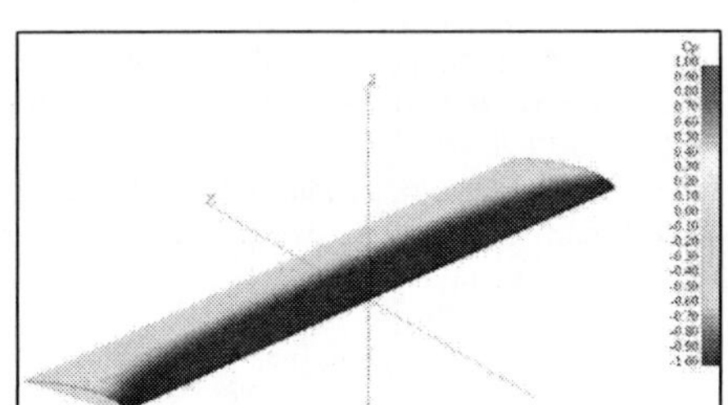

(b) Contours of coefficient of pressure

Fig. 7 Flow visualization by using panel method

In experiments, it is clear that the C_{L_α} (per degree) is high in comparison with those obtained by using the panel method. Its experimental value is 0.0913 and analytical value is 0.0617. Based on these results; it is proposed to apply a correction factor of 1.36 to the analytical value. Moreover, the value of C_{L_α} obtained by using (2) is 0.09 (/deg). C_{L_α} value predicted by using the above mentioned relationship is perhaps same as that predicted by the experiment. It is important to note that (2) will provide the value of C_{L_α} in per radian and for the purpose of comparison of results, that digit has to be then converted into per degree.

$$C_{L_\alpha} = \frac{2\pi A_R}{2+\sqrt{4+A_R}} \quad (2)$$

Irrespective of the difference between the true Reynolds number and that achievable in the wind tunnel, the results of C_{L_α} obtained from this study will be helpful for applying the C_{L_α} with correction factor for the aerodynamic and stability analysis of a canard based model of a HB vehicle. Moreover, the effect of the interaction between the model and the sidewall boundary layers for the model is unknown, which may affect the quality of the comparison between predictions and measurements for this configuration. In this regard, future work will include the study of the effect of the gap/standoff height, like keeping it equal to the boundary thickness where the local velocity is 0.9 times the free steam velocity and look into the overall behavior of the aerodynamic coefficients.

V. Conclusion

Wind tunnel testing was carried out to assess the aerodynamic characteristics of canard of a hybrid buoyant aircraft. In comparison with the experimental data; VLM code underestimates the lift coefficient. This loss of lift increases with incidence, resulting in a reduced lift curve slope. A correction factor is proposed for accurate prediction of the same by using VLM. The comparative study confirms the existence of the relationship of standoff height to be equal to the boundary layer displacement.

Acknowledgment

The supports of the Ministry of Science, Technology and Innovation (MOSTI), Malaysia, under the grant 06-01-08-SF0189 and the Ministry of Education, Malaysia under the grant FRGS 13-020-0261 are gratefully acknowledged. Authors are thankful to Azliza Binti Embong, Senior Engineer IIUM-LSWT, and her team for providing assistance during the wind tunnel experimental work.

References

[1] E. Reshotko and W. S. Saric, "Flow quality issues for large wind tunnels Flow Quality Issues for Large Wind Tunnels," AIAA Paper (1997): 97-0225.

[2] D. Kiichemann, "Problems in Wind Tunnel Testing Techniques." AGARD Report No. 601, 1973

[3] A. U. Haque, W. Asrar, A. A. Omar, E. Sulaeman, and J. S. Mohamed, "Assessment of Engine ' s Power Budget for Hydrogen Powered Hybrid Buoyant Aircraft," Journal of Power and Propulsion, 2015 (in press).

[4] A. U. Haque, W. Asrar, A. A. Omar, E. Sulaeman, and M. A. JS, "Conceptual Design and Sizing of a Winged Hybrid Airship," 21st AIAA Lighter-Than-Air Systems Technology Conference, 16-20 June 2014, Atlanta, GA. Paper No. AIAA–2014–2710.

[5] J. T. Tambahan and S. D. Ehsan, "winglet technology for the modern aircraft In Proceedings of the 3rd BSME-ASME International Conference on Thermal Engineering (Vol. 20, p. 22).

[6] R. Hills, "A Review of measurements on AGARD calibration models,"Agardograph-64, Advisory Group For Aeronautical Research And Development Paris (France), 1961.

[7] N. Gross and M. R. Number, "The ETW wall interference assessment for full and half models, AIAA paper, 769, 2004.

[8] J. F. Marchman, W. J. Devenport, W. H. Mason, and T. Experiments, "Gap Size Effect on Low Reynolds Number Wind Tunnel Experiments. (Doctoral dissertation, Virginia Polytechnic Institute and State University), 1999.

[9] A. Group, F. O. R. Aerospace, and A. A. Report, A Selection of Experimental Test Cases for the Validation of CFD Codes, AGARD AR, 2(303), 1994 .

[10] W. H. Rae, & A. Pope (1984). Low-speed wind tunnel testing. John Wiley.

[11] H. C. Garner, E. W. Rogers, W. E. Acum, & E. C. Maskell, Subsonic wind tunnel wall corrections (Agardograph-109), 1966, Advisory Group For Aerospace Research And Development Neuilly-Sur-Seine (France).

[12] A. Malik, "Suppression of junction flow effects in half model wind tunnel testing." PhD Thesis, Loughbergh Univerity, UK, 2011

[13] M. R. Soltani, A. Mamaghani, and A. Bakhshalipour, "Half-Model Testing and Sidewall Effects," 25th International Congress of the Aeronautical Sciences, 3 - 8 September 2006, Hamburg, Germany

[14] S. Kuppa and Marchman III, J. F. End plate gap effects on a half wing model at low Reynolds nUfy, bers. In a Collection of Technical Papers: AIAA 5th Applied Aerodynamics Conference, August 17-19, 1987, Monterey, California, American Institute of Aeronautics and Astronautics.

[15] H. Schlichting and K. Gersten, Schlichting, H., Gersten, K., & Gersten, K., Boundary-layer theory. Springer Science & Business Media, 2000.

[16] A. Deperrois, XFLR5: a tool for the design of airfoils, wings and planes operating at low Reynolds numbers. Software Package.", 2010

[17] S. G. Hedman, Vortex lattice method for calculation of quasi steady state loadings on thin elastic wings in subsonic flow (No. FFA-105). Aeronautical Research Inst of Sweden Stockholm, 1966.

Conceptual Design of Unmanned Aerial Targets

M. Adamski, J. Cwiklak

***Abstract*—**The contemporary battlefield creates a demand for more costly and highly advanced munitions. Training personnel responsible for operations as well as immediate execution of combat tasks which engage real asset is unrealistic and economically not feasible. Owing to a wide array of exploited simulators and various types of imitators, it is possible to reduce the costs. One of the effective elements of training, which can be applied in the training of all service branches, is imitator of aerial targets. This research serves as an introduction to the commencement of design analysis over a real aerial target imitator. Within the project, the basic aerodynamic calculations were made, which enabled to determine its geometry, design layout, performance as well as mass balance of individual components. The conducted calculations of the parameters of flight characteristics come closer to the real performance of such Unmanned Aerial Vehicles.

***Keywords*—**Aerial target, aerodynamics, imitator, performance.

I. Introduction

THE use of more costly and technologically advanced weapons and equipment in contemporary armed forces creates a demand to modify the training process of military personnel. Troops should practise in conditions which are as close as possible to real conditions in theatre. The procurement and maintenance costs of modern armaments are steadily growing. One of the ways of reducing these expenditures is to rationalize the training expenses. Owing to a common application of simulators and various types of imitators, it is possible to lower the costs, both in the training of land forces, crews of Unmanned Aerial Vehicles, naval ships and the air defence. One of the most effective training elements, exploited in the above-mentioned services, are imitators of aerial targets.

The low cost of manufacturing an imitator makes its loss in combat operations not so painful as in the case of other aerial vehicles. The cost of shooting down such an imitator may prove more expensive than the value of the imitator itself. Imitators of aerial targets may successfully be used in deceiving the enemy air defence. Simulating air assaults by means of the imitators may lead to lowering the enemy combat readiness, due to the necessity of engaging its elements to counteract a potential threat [4]. The capabilities of using imitators of aerial targets also include airstrikes. An imitator equipped with munitions is a target which is dangerous and difficult to eliminate. The cost of obtaining it is considerably lower than the value of potential targets which may be destroyed. A broad spectrum of applying this sort of unmanned vehicles and a relatively low manufacture cost, compared with other aerial vehicles, renders exploitation of suchlike drones justifiable, in the armed forces [1].

M. Adamski and J. Cwiklak are with the Polish Air Force Academy, Dywizjonu 303/35, 08-530 Deblin, Poland (corresponding author: M. Adamski; phone: +48 815517423; e-mail: j.cwiklak@wsosp.pl, terazen@wp.pl).

II. Project Assumptions

A preliminary design of an Unmanned Aerial Vehicle should include tactical, technical, technological and exploitation needs. However, meeting all these requirements is often connected with a great deal of difficulties, since they are often contradictory and in many cases impossible to combine [9].

Basic geometrical dimensions:

- wing span – l = 3 m;
- length – l_k = 2.60 m;
- maximum length of fuselage – 0.4 m;
- maximum height of fuselage – 0.4 m;
- length of root chord – b_n=0.56 m;
- length of tip chord – b_k=0.2 m;
- wing surface area- S=1.14 m^2.

Other assumptions:

- cruising speed -350 km/h;
- propulsion – jet engine;
- ceiling – 6000 m;
- Maximum Take-off Mass – 90 kg.

Judging by the development tendencies of the exploited imitators (measurements, types of propulsion, tasks and the like), basic geometrical dimensions and aerodynamics as well as the initial assumptions for flight parameters were specified [2].

III. Model Mass

The knowledge of particular components and other major elements of an aircraft, as well as the assumed payload, which add to the weight during a flight, is essential to determine the location of the centre of gravity, and to establish its approximate weight [3].

Notation used in the calculations:

- m_s – mass of wing;
- m_{to} – aircraft take-off mass;
- n_A – coefficient of maximum permissible payload;
- υ – factor of safety;
- Λ – aspect ratio;
- g_o – relative thickness of root airfoil;
- χ – wing sweep of the quarter chord line (0.25);
- S – lift surface;
- τ – taper ratio;
- φ – load factor;
- Δp_k – maximum pressure differential limit in fuselage;
- V_{cruise} – cruising speed;
- q_k – unit mass of fuselage;

- S_k – fuselage front surface;
- S_{omk} –fuselage area of airflow;
- D_k - hydraulic diameter of fuselage.

On the basis of the calculations, it is assumed that the mass of the wing equals $m_s \approx 6$ kg and mass of fuselage is assumed to equal $m_k \approx 9$ kg. Due to the use of V-tail, the surface area of an elevator unit and a vertical tail unit have been replaced with projections of these planes on the horizontal and vertical plane, respectively. Taking into account the above results, the mass of tail is assumed to be equal to $m_{hv} \approx 2.2$ kg.

TABLE I
LISTING OF MASSES

Lp.	Component	Mass [kg]	Distance [m]
1	Radar reflector	$m_1 = 1$	$x_1 = 0.127$
2	Cargo area	$m_2 = 3$	$x_2 = 0.379$
3	Autopilot+GPS+INS	$m_3 = 6$	$x_3 = 0.81$
4	Cargo area	$m_4 = 7$	$x_4 = 0.81$
5	MDI	$m_5 = 5.89$	$x_5 = 1.26$
6	Fuselage	$m_6 = 9$	$x_6 = 1.26$
7	Fuel tank	$m_7 = 31.5$	$x_7 = 1.426$
8	Wing	$m_8 = 6$	$x_8 = 1.438$
9	Pylons	$m_9 = 2$	$x_9 = 1.438$
10	Parachute	$m_{10} = 5$	$x_{10} = 1.9$
11	Batteries	$m_{11} = 3$	$x_{11} = 1.867$
12	Steering system	$m_{12} = 1.3$	$x_{12} = 2.1$
13	Flare dispenser	$m_{13} = 3$	$x_{13} = 2.2$
14	Control surfaces	$m_{14} = 2.2$	$x_{14} = 2.33$
15	Engine	$m_{15} = 4.1$	$x_{15} = 2.33$
	TOTAL $\sum$	$m_{to} = 90$	$x_{sa} = 21.472$

On the basis of the data listed in Table I and (1), it is possible to calculate the position of the centre of gravity of the designed imitator of an aerial target.

$$x_{sa} = \sum_{n=1}^{n=15} \frac{x_n \cdot m_n}{m_{to}} \ [m] \tag{1}$$

The centre of gravity against the longitudinal axis is away from the nose edge of the designed aerial vehicle by x_{sa}=1.41m, which is included within 0.25-0.33 mean aerodynamic chord of the wing.

IV. AIRFOIL

Taking into account the geometric data and the analysis of available aerial vehicles of this type, the value of Reynolds number has been calculated, which is equal to the assumed minimum steady flight speed of 100 km/h (at low levels) [2]:

$$R_e \frac{Vsl \ \times b_{sa}}{V_0} \tag{2}$$

where: Re – Reynolds number; V_{sl} =27.7 m/s - velocity; b_{sa} =0.4084 m – length of mean aerodynamic chord; ν_0 =0.0000143 m^2/s – coefficient of kinematic viscosity. For the above data Re≈$0.8 \cdot 10^6$.

Taking advantage of the X-foil-5 programme, from the database of the available airfoils, the nine 6-series low-drag airfoils were selected. They were next subject to simulation with the panel method. The obtained characteristics were z=f(α) and cx=f(α) for Reynolds number Re=$0.8 \cdot 10^6$.

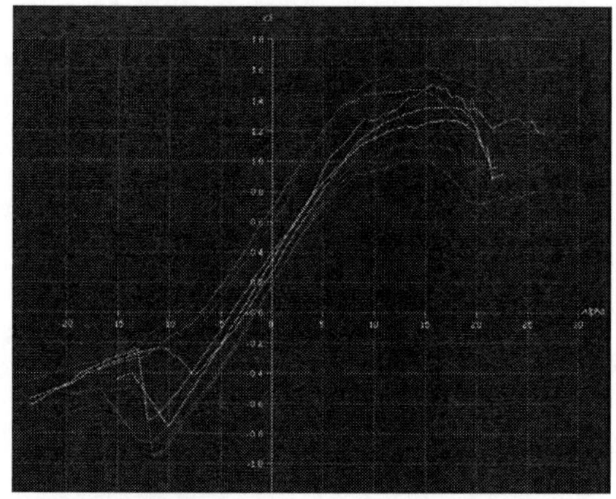

Fig. 1 Characteristics of selected airfoils c_z=f(α)

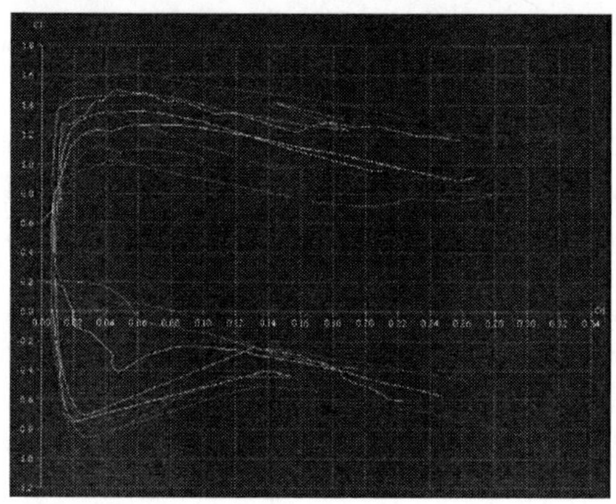

Fig. 2 Characteristics of selected airfoils c_x=f(α)

On the basis of the obtained graphs, the airfoil NACA $63_{(2)}$215B was selected, which is characterized by low drag and a wide range of useful angles of attack.

Using the data from the X-foil 5 programme, for Reynolds number $0.8 \cdot 10^6$, the following airfoil characteristics of c_z=f(α) and c_x=f(α) have been generated:

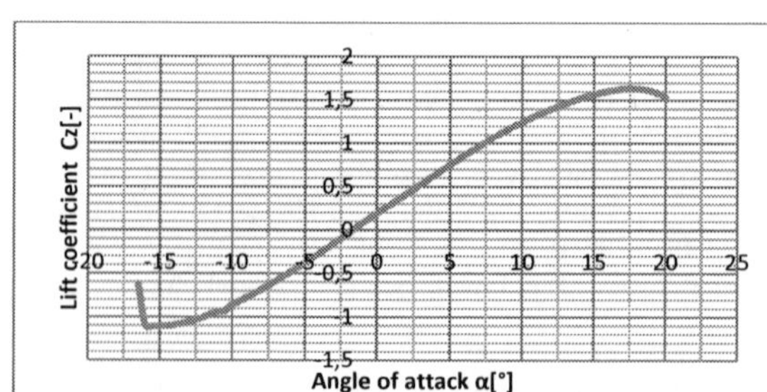

Fig. 3 Dependence of the lift coefficient on the angle of attack

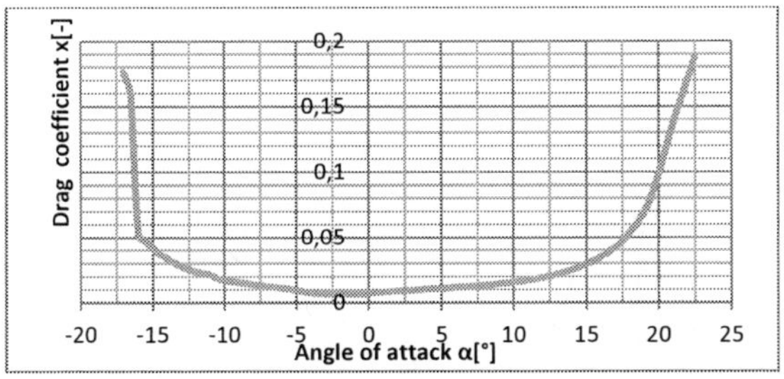

Fig. 4 Dependence of the drag coefficient on the angle of attack

V. Determining Drag Coefficient of Particular Elements

The main element which affects the change in drag, depending upon the lift, is a three-dimensional airflow of the wing. In the subsonic range, wing drag can be shown depending upon the drag which occurs in two-dimensional flow and drag induced as follows [3]:

$$c_{xs} = c_{xmin} + \frac{c_z^2}{\pi \cdot \Lambda_e} \quad (3)$$

Summing up the particular values and using the formula for aspect ratio, value of the aspect ratio Λ_e=7.178 was obtained, which entails the value of the wing drag coefficient c_{xs}=0.0788.

Fuselage drag coefficient, similarly to the wing drag, for currently used well-shaped fuselages, for the Ma number below the critical value, can be expressed as:

$$c_{xk} = c_f \cdot \eta_k \cdot \frac{S_{ck}}{S_k} + \Delta c_{xm} \quad (4)$$

where: c_f – drag coefficient of fuselage friction, depending upon the type of flow; η_k – coefficient allowing for the impact of fuselage thickness; η_{Ma} – coefficient allowing for compressibility; S_k – largest cross-sectional area of fuselage; S_{ck} – total fuselage surface in contact with airflow; Δc_{xm} – interference drag resulting from air intakes, which affect the fuselage.

Taking into consideration the fact that the designed imitator is characterized by not very high airspeeds, the value of the coefficient $\boldsymbol{\eta_{Ma}}$ is taken as equal to 1.

Summing up the above deliberations, the final value of the fuselage drag is obtained from the dependency 4.2 c_{xk} =0.062.

On the basis of the above values of drag coefficients for various values of the lift coefficient, it is possible to determine the polar curve of an aerial vehicle.

The drag coefficient for the whole unmanned vehicle has been calculated from:

$$c_x = \frac{c_{xs} \cdot S + c_{xk} \cdot S_k + c_{xu} \cdot S_u}{S} \cdot (1 + K_{int}) \quad (5)$$

where the interference coefficient is rather different for various aircraft, and even for different c_z.

It may be assumed that K_{int} ranges from 0.05 to 0.07. For the sake of calculations, it is assumed that K_{int}=0.05.

For c_z=1.2512, the total drag equals: c_x=0.124. On the basis of the above data, the polar curves for the whole wing and imitator were obtained.

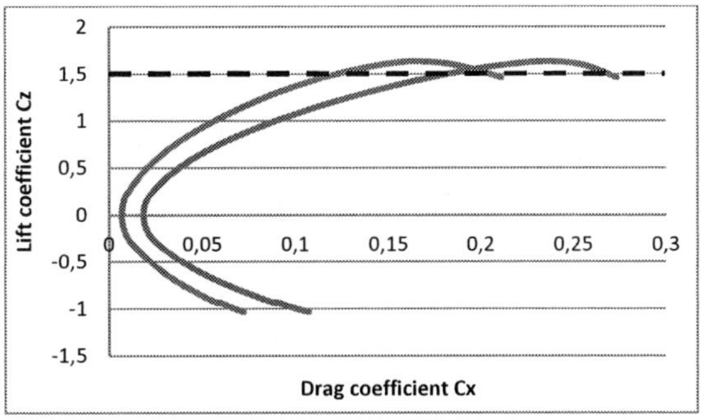

Fig. 5 Characteristics c_z=f(c_x)

VI. Section of a Propulsion System

The selection of the engine type is determined by the assumed airspeeds which an aerial vehicle is to achieve, as well as such requirements as: economy, repair and maintenance [6]. The design at stake assumes that it will be a jet engine, placed in the rear part of the fuselage, manufactured by the Safran group company - Microturbo TR-3, 300 N of thrust.

Due to lack of available characteristics of the engine, for calculations, we used a mathematical model; the thrust dependence on the Mach number has been obtained from the dependence:

$$P_r = P_{r0} \cdot \eta_{Ma} \cdot \eta_H \cdot \eta_{\bar{n}} \quad (6)$$

where: P_{r0} – engine thrust at the altitude h=0[m], in normal conditions; η_{Ma} – coefficient allowing for the influence of the Mach number; η_H – coefficient allowing for the influence of the flight altitude; $\eta_{\bar{n}}$ – coefficient allowing for the turbine rotations - $\eta_{\bar{n}} = 1$

$$\eta_{Ma} = (1 - 0{,}1Ma \cdot s_{dw}) \cdot (1 - 0{,}5Ma + 0{,}6Ma^2) \quad (7)$$

for h<11000 m

$$\eta_H = {\sigma_H}^{0{,}95 - 0{,}21Ma^2} \quad (8)$$

where: σ_H – ratio of air density at h altitude versus density at density h=0 m; s_{dw} – dual circuit (range) – s_{dw}=1.

VII. Determining Basic Performance

The type of a selected method in order to determine the basic performance is affected by the so-called value of thrust indicator) $\mathbf{q_{Pr}}$ defined as:

$$q_{Pr} = \frac{m_{to}}{P_{R-max}} \quad (9)$$

where: m_{to}=90 kg; P_{R-max} =0.3 kN

Since the thrust indicator equals: q_{Pr}=300 kg/kN, in order to calculate the performance, we can use the power method [11].

The obtained dependencies enable to determine the dependencies of rate of climb and the angle of the flight path upon an altitude and a flight speed.

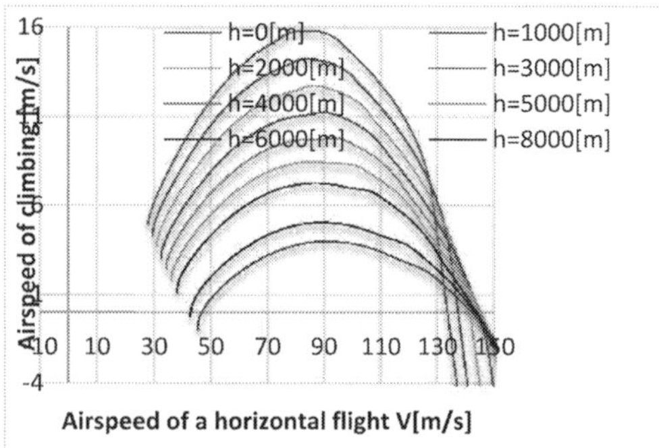

Fig. 6 Dependence of rate of climb on airspeeds for various altitudes

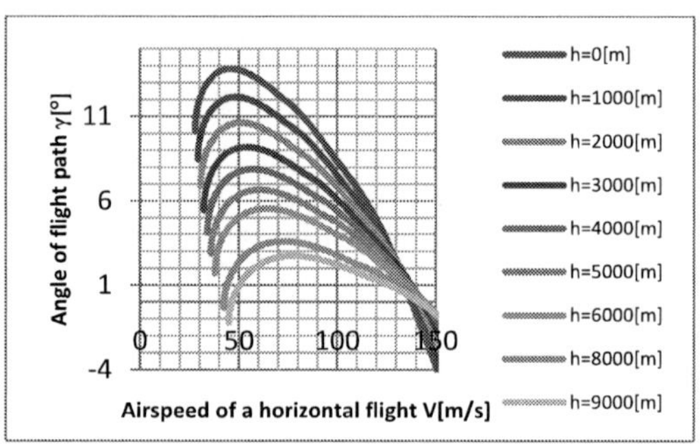

Fig. 7 Dependence between the angle of a flight path and airspeed for various altitudes

TABLE II
UAV PERFORMANCES IN RELATION TO FLIGHT ALTITUDE

h [m]	V_{min} [km/h]	V_{max} [km/h]	w_{max} [m/s]	V_w [km/h]	γ_{max} [°]	V_γ [km/h]
0	100.26	480.2	15.8	290.9	13.82	168
1000	105.25	493.2	14.22	305.3	12.15	176
2000	110.61	508.5	12.68	320.9	10.63	184.86
3000	116.38	521.28	11.1	324	9.2	194.5
4000	122.61	527	9.74	316	7.9	205
5000	129.31	528.12	8.47	317	6.67	216.14
6000	136.62	523.8	7.29	318	5.56	228.35
8000	153.44	514.8	5.04	320	3.58	256
9000	162.54	514	4	322.6	2.72	271.62

In conclusion to the above deliberations, on Fig. 8, it can be depicted basic performance of an aerial vehicle as functions of a flight altitude [10]. It also contains, apart from the previously calculated values, the service ceiling, and the absolute ceiling, which due to the limitations of the exploited engine equals 9000 m. At this altitude, the maximum rate of climb equals W_{max} = 4 m/s at the flight speed of V=322.6 km/h.

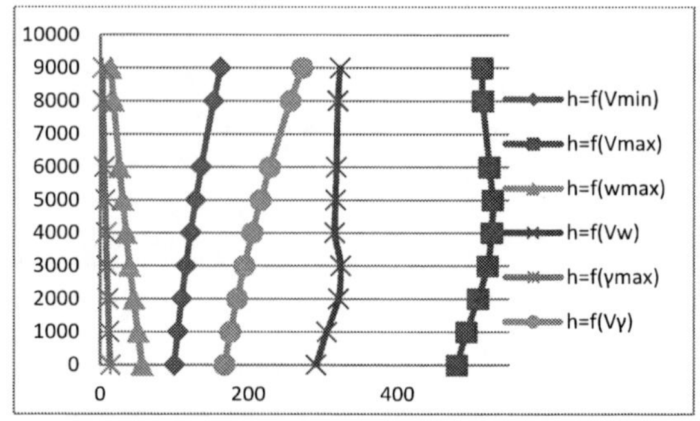

Fig. 8 Performance of the designed imitator of an aerial target

The main idea of designed imitator of an aerial target is to train the air defence personnel as well as pilots in fighting off air targets. The mission profile must meet the requirements connected with launching air defence missiles and also the requirements connected with eliminating air targets with fighter aircraft [7], [8].

The Unmanned Aerial Vehicle takes off on a launch pad. Next, the imitator of an air target, executes a task in the autonomous mode, in accordance with the programmed profile and a flight path over a designated area. The flight profile is uploaded each time before the takeoff. The flight path tracking is performed by a ground control station, which is also capable of changing the parameters during a mission. On completion of the task, a parachute set is activated for a safe landing [5].

VIII. CONCLUSION

Designing an aerial vehicle is a complex process, which covers actions that are analytical, technical, and organizational in their character. The major design work involves large teams and high financial expenditures. The costs of research and development exceed the price of one aerial vehicle by several times. Therefore, it is understandable that these stages of design which significantly influence the final output receive particular attention. Long-term experience of various companies proves that designing which is supported by extensive analysis of current needs brings about best results.

The adopted concept of the imitator of an aerial target has been shaped on the basis of a scrutiny of trends of models which are currently used. Within the project, the basic aerodynamic calculations had been made, which have enabled to determine performance and mass balance of particular components. Owing to the calculations, it was possible to bring closer the flight and technical parameters to the real performance of Unmanned Aerial Vehicles [8]. This investigation is an introduction to a design analysis over an imitator of an aerial target. Aerial vehicles of this kind are a considerable help in combat training of air units and air defense units. A low cost of the manufacture of the imitator, compared to other aerial vehicles, justifies conducting design work. Potential possibilities of exploiting imitators of aerial targets cause that its use is fully justified from the economic and military standpoint.

References

[1] Adamski M., Rajchel J., „Bezzałogowe Statki Powietrzne, cz. I, Charakterystyka i wykorzystanie", WSOSP-Dęblin, 2013.

[2] Danielecki S., Projektowanie samolotów, Oficyna Wydawnicza Politechniki Warszawskiej, Warszawa, 2000.

[3] Fiszdon W., Mechanika lotu , PWN, Warszawa – Łódź, 1961.

[4] Grzesik N., Infrared passive homing guidance gun system on UCAV, Scientific Aspects of Unmanned Mobile Vehicle, Politechnika Świętokrzyska, Kielce, 2010.

[5] Grzegorzewski M., Ciećko A., Oszczak S., Ćwiklak J., Aircraft Landing System Utilizing a GPS Receiver with Position Prediction Capability/Functionality, ION GNSS 2011. Portland USA.

[6] Jankowska-Sieminska B., Jankowski A., Slezak M., Analysis and Research of Piston Working Conditions of Combustion Engine in High Thermal Load Conditions, Journal f KONES 2007, Vol. 14, No. 3.

[7] Jel R., Projekt wstępny imitatora celu powietrznego., Praca mgr WSOSP-Dęblin, 2010.

[8] Koruba Z., Dziopa Z., Krzysztofik I., Dynamics of a Controlled Anti-aircraft Missile Launcher Mounted on a Moveable Base, Journal of Theoretical and Applied Mechanics, Vol. 48, No. 2, 2010.

[9] Pila J., Kozuba J., Korba P., Airframe Structure, Dęblin, 2014.

[10] Warchulski J., Czechowicz B., Opracowanie projektu systemu imitatorów celów powietrznych do strzelań rakietowych ICP-SR, Warszawa, 2007.

[11] Zyluk A., Numerical simulation of the Effect of Wind on the Missile Motion, Journal of Theoretical and Applied Mechanics, 52/2, 2014.

Wind Tunnel for Aerodynamic Development Testing

E. T. L. Cõuras Ford, V. A. C. Vale, J. U. L. Mendes, F. A. Ribeiro

***Abstract*—**The study of the aerodynamics related to the improvement in the acting of airplanes and automobiles with the objective of being reduced the effect of the attrition of the air on structures, providing larger speeds and smaller consumption of fuel. The application of the knowledge of the aerodynamics not more limits to the aeronautical and automobile industries. Therefore, this research aims to design and construction of a wind tunnel to perform aerodynamic analysis in bodies of cars, seeking greater efficiency. Therefore, this research aims to design and construction of a wind tunnel to perform aerodynamic analysis in bodies of cars, seeking greater efficiency. For this, a methodology for wind tunnel type selection is designed to be built, taking into account the various existing configurations in which chose to build an open circuit tunnel, due to the lower complexity of construction and installation; operational simplicity and low cost. The guidelines for the project were teaching: the layer that limits study and analyze specimens with different geometries. For the variation of pressure in the test, section of a switched gauge used a pitot tube. Thus, it was possible to obtain quantitative and qualitative results, which proved to be satisfactory.

***Keywords*—**Wind tunnel, Aerodynamics, Air.

I. Introduction

The effect of the air if moving around a vehicle effects of three different manners your behavior. These three manners are:

- Resistance to the movement;
- Sustentation effects;
- Effect of lateral winds.

The builders' first concern was exactly with the problem of the aerodynamic resistance, since this affects the potency sensibly consumed by the vehicle. Although the first detailed studies have been initiate in 1920, until today most of the cars possesses a form that takes the waste of potency of the order from 30 to 40%. With relationship to the lateral winds, the project of the drag of the models in production almost disrespects completely them [1]-[5]. These facts are due mainly the following causes:

- Almost completely aerodynamic study should be experimental or numeric, with great expenditure of time and money;
- The best solutions in terms of aerodynamics adapt badly to an automobile project, in design terms and internal space [1].

The study of the forces generated on submerged objects in a draught it is applied to the development and optimization of projects and products in areas as sport, ergonomics, industrial design and it would engineer mechanics and civil. The aerodynamic study can be accomplished being used tunnels of wind of different types, which can be classified in function of the construction position (vertical or horizontal), of the type of the camera (open or closed) and of the speed of the fluid (subsonic or hypersonic). In general, a convergent mouthpiece, a test section and a fan compose a wind tunnel. The convergent mouthpiece is a fundamental component for an aerodynamic evaluation with smaller involved uncertainties. According to [2]-[9] the principal effects of the mouthpiece are the reduction of the flotation of speed in the test, section and the increase of the medium speed of the drainage. The test section requests cares in the production and assembly allowing the drainage to be as uniform as it is possible, it exempts of angles and with the minimum of whirls [10]-[17]. This study proposes the project and the construction of a wind tunnel for the accomplishment of aerodynamic rehearsals seeking the best understanding of the interaction between the fluid and the structure, on the part of the academics of the course of mechanical engineering of the Federal University of Rio Grande do Norte.

II. Materials and Methods

To simulate the effect of the air passing on the surface of the trucks of the vehicles, it was necessary the construction of an aerodynamic tunnel.

The tunnel possesses the following characteristics: total length of 4,5m; rehearsal section with dimensions of 0,50m X 0,50m, with 0,25m2 area; maximum speed of approximately 9,7m/s, obtained using a monophasic electric motor, with maximum potency of 5HP (3.7kW). Diffuser with reason of contraction of 2:1, in other words, with larger section of 1m X 1m, and section smaller 0,50m X 0,50m, that according to [4], does with that flowed him when passing for that area has an increase of the speed due to decrease of the pressure. Figs. 1-6 and Table I detail the construction and assembly of the tunnel [18]-[20].

Elmo Thiago Lins Cõuras Ford, is Ph.D in Mechanical Engineering; M.Sc. in Administration; M.Sc. in Mechanical Engineering; Engineer of Safety; Mechanical Engineer; Automotive Engineer. Universidade Federal do Rio Grande do Norte – UFRN / PPGCEM, Av. Senador Salgado Filho, S/N, 59072-970, Natal - RN, Brazil (e-mail: elmocouras@hotmail.com).

Valentina Alessandra Carvalho do Vale, is Ph.D. Student in Electrical Engineering; M.Sc. in Mechanical Engineering; Engineer of Safety; Electrical Engineer. Universidade Federal do Pernambuco – UFPE / PPGEE, Recife – PE, Brazil. (e-mail: ale.vale.ssandra@gmail.com).

José Ubiragi Lima Mendes, is Ph.D. in Science and Engineering of Materials; M.Sc. in Mechanical Engineering; Mechanical Engineer. Universidade Federal do Rio Grande do Norte – UFRN / PPGEM, Av. Senador Salgado Filho, S/N, 59072-970, Natal - RN, Brazil (e-mail: ubiragi@ct.ufrn.br).

Fernanda Alves Ribeiro, is Ph.D. Student in Mechanical Engineering; M.Sc. in Mechanical Engineering; Mechanical Engineer. Universidade Federal do Rio Grande do Norte – UFRN / PPGEM, Av. Senador Salgado Filho, S/N, 59072-970, Natal - RN, Brazil (e-mail: ribeiro.f.a.@hotmail.com).

Fig. 1 Molds for the diffuser

Fig. 2 First built piece

Fig. 3 Assembly of the contraction

Fig. 4 Armed diffuser

Fig. 5 Construction of the section of rehearsals

Fig. 6 Centrifugal fan

TABLE I
SIZING THE NUMBER OF COLLECTORS

Length	4,5 m
Maximum Height	1 m
Maximum Width	1 m
Type of the Tunnel	Soprador
Maximum Potency of the Motor	5 HP (3,7 kW)
Area of the Rehearsal Section	0,25m^2
Maximum Speed of the Rehearsal Section	9,7m/s

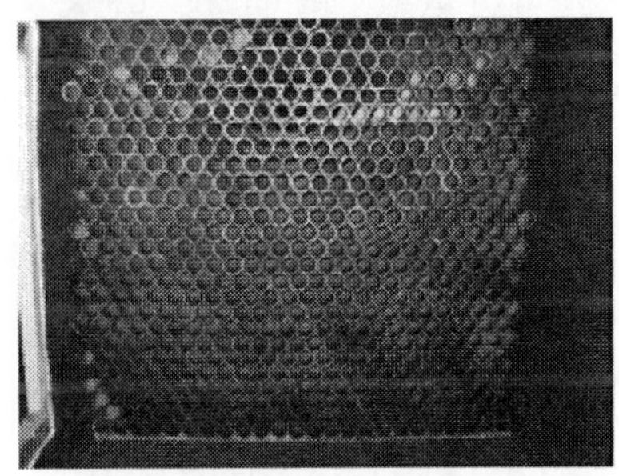

Fig. 7 Box of harmonization of the type beehive

In the tunnel of used wind, a centrifugal fan, moved by a motor alternating current tramway, blows the air. The air goes by a diffuser in which there is in your exit extremity to a box of the type beehive. It was also projected and built a camera of stabilization with a rectification of flow of the type beehive. They considered in the dimension of the camera of stabilization the observations of [8], according to which the length of the honeycomb should be from six to eight times the medium diameter of the cell, and the recommendation of [6], that stipulates a strip for the thickness of the wall of the cells between 0.5 and 2.0mm. The beehive was built starting from tubes of PVC that were cut with 200mm of length each one. Those parts were mounted and agglutinated, forming 625 cells of square section 50x50mm. In Fig. 7, the final assembly of the beehive can be observed.

After the diffuser, the drainage suffers an acceleration, due to the presence of a contraction in which the drainage suffers a pressure loss, that is, turns into kinetic energy (speed) that goes to the rehearsal section. The whole construction of the structures of the tunnel (diffuser, contraction, rehearsal section), they were accomplished by techniques used by [4]. Parts as: The box of harmonization of the drainage; the cooker hood to retain the coming vibrations of the motor and the helix of the motor for generation of the wind, technical were accomplished used at the Laboratory of Aerodynamics of Vehicles of the Lutheran University of Brazil - ULBRA. Figs. 8 and 9 show a view of the Pitot tube.

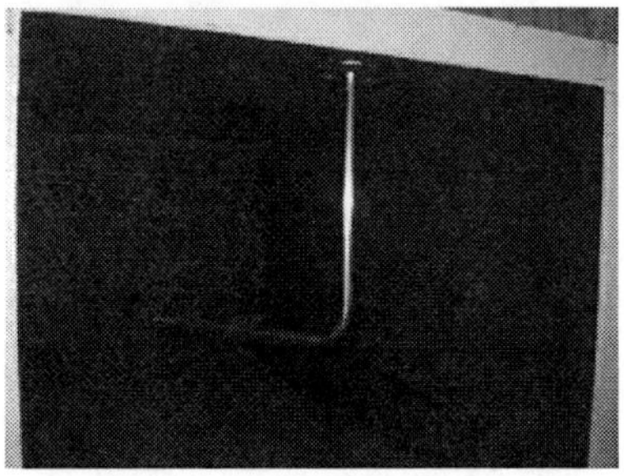

Fig. 8 Pitot tube

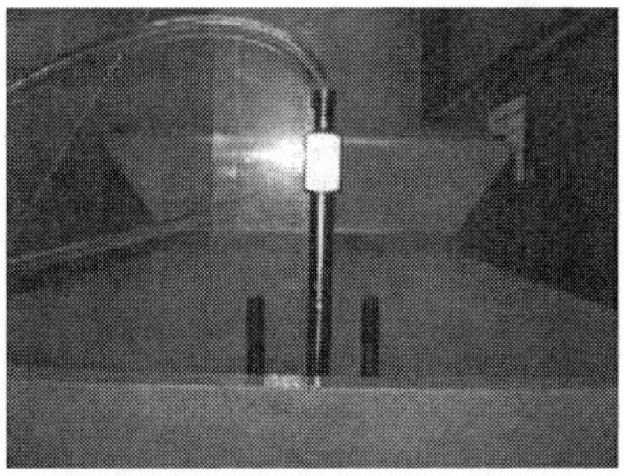

Fig. 9 Height adjustment of the Pitot tube

III. Results and Discussion

In Fig. 10, the tunnel of projected wind is presented. However, the project stage and construction is in final phase, as aerodynamic scale and system of smoke for visualization, for the longed for didactic practices.

An important result was obtained starting from the star-up of the tunnel. It treated initially of the inclusion need to the project foreseen of the expansion camera with the harmonization beehive, reason of the rotation effect imposed to the drainage by the axial fan. For it tests them initials objects of different formats they were used, chosen in a random way.

Fig. 10 Tunnel of installed wind

IV. Conclusion

Before the rehearsals accomplished in elapsing of this research, the wind tunnel behaved in a satisfactory way, having reached the objective for which was built, being this of easy construction, simplified operation and of low cost. The expansion camera with beehive was fundamental to minimize the effect of the rotation imposed to the drainage by the axial fan.

Acknowledgment

Thanks go to Coordenação de Pessoal de Nível Superior - CAPES, Conselho Nacional de Desenvolvimento Científico e Tecnológico – CNPQ, Universidade Federal de Pernambuco – UFPE, Universidade Federal do Rio Grande do Norte - UFRN and ULBRA, for providing the means to carry out this work.

References

[1] 1998. Aerodynamics of road vehicles: from fluid mechanics to vehicle engineering / Syed R. Ahmed.; edited by Wolf-Heinrich Hucho. Warrendale: SAE, 918 p.
[2] Bell, J. H., Mehta, R. D. 1989. Boundary-layer predictions for small low speed contractions American Institute of aeronautics and Astronautics, v.27, n.3, p. 372-174.
[3] Canale, A. 1989. Automobilística: Dinâmica e Desempenho. Ed. Érika: São Paulo.
[4] Catalano, F. M. 1988. Projeto, construção e calibração em túnel aerodinâmico de circuito aberto tipo N.P.L. de seção transversal hexagonal. USP: São Carlos.
[5] Carregari, A. L. 2006. Estudo do escoamento de ar sobre a carroceria de um ônibus usando programa de CFD e comparação com dados experimentais. USP: São Carlos.
[6] Groff, J. E. A.; Alé, J. V. 2000. Projeto de túnel de vento subsônico de circuito aberto. Rio Grande do Sul: PUC.
[7] Fox, R. W. 2001. Introdução à mecânica dos fluidos. McDonald, LTC: Rio de Janeiro 504p.
[8] Laranja, R. A. C. 2005. Mecânica dos fluidos. Ed. ULBRA: Canoas. 87p.
[9] Ahmed, S. R. Wake Structure of typical automobile shapes. Journal of Fluids Engineering, Vol. 103, Mar 1981.
[10] Ahmed, S. R. Influence of base slant on the wake structure and drag of road vehicles. Journal of Fluids Engineering, Vol. 105, Dec. 1983. ABNT. 1997. Resíduos Sólidos: ABNT/NBR 10004. Rio de Janeiro.
[11] Anderson, J. D. Fundamentals of Aerodinamics. 3 ed., McGraw-Hill, 2001.
[12] Câmara, E. F. Estudo aerodinâmico do escoamento sobre modelo de ônibus. ITA: São José dos Campos, 1993.
[13] Campbell, C. The sports car. Its design and performance. UK: Chapman and Hall Ltda. 1969.

[14] Diuzet, M. The moving-belt of the I.A.T. Long test section wind tunnel. Journal of Wind Engineering and Industrial Aerodynamics, Vol. 22, pp. 237-244, 1986.
[15] Hucho, W. H. Aerodynamics of Road Vehicles. Butterworth, 1987.
[16] Kline, S. J.; McClintock, F.A. Describing Uncertainties in single-sample experiments. Mechanical Engineering, Jan. 1953.
[17] Lajos, T.; Preszler, L. Effect of moving ground simulation on the flow past bus models. Journal of Wind Engineering and Industrial Aerodynamics, Vol. 22, pp. 271-277, 1986.
[18] Nicolazzi, L. C; Rosa, E. Introdução à modelagem de veículos automotores. GRANTE-UFSC: Florianópolis, 2001.
[19] Videira, L. C. P. Estudo Experimental de um modelo de ônibus com ênfase no escoamento da esteira. ITA: São José dos Campos, 2001.
[20] Cõuras Ford, E. T. L. Análise quali-quantitativa do deslocamento da camada limite em carrocerias veiculares. UFRN, 2008.

Aerodynamic Study of Vehicle Wind Tunnel and Water Tunnel for Analysis of Bodies

E. T. L. Cöuras Ford, V. A. C. Vale, J. U. L. Mendes

***Abstract*—**The simulation in wind tunnel is used thoroughly to model real situations of drainages of air. Besides the automotive industry, a great number of applications can be numbered: dispersion of pollutant, studies of pedestrians' comfort, and dispersion of particles. This work had the objective of visualizing the characteristics aerodynamics of two automobiles in different ways. To accomplish that drainage of air a fan that generated a speed exists (measured with anemometer of hot thread) of 4,1m/s and 4,95m/s. To visualize the path of the air through the cars, in the wind tunnel, smoke was used, obtained with it burns of vegetable oil. For "to do smoke" vegetable oil was used, that was burned for a tension of 20V generated by a thread of 2,5mm. The cars were placed inside of the wind tunnel with the drainage of "air-smoke" and photographed, registering like this the path lines around them, in the 3 different speeds.

***Keywords*—**Aerodynamics, Vehicle Drag, Wind tunnel.

I. Introduction

THE effect of the air if moving around a vehicle effects of three different manners your behavior. These three manners are:

- Resistance to the movement;
- Sustentation effects;
- Effect of lateral winds.

The builders' first concern was exactly with the problem of the aerodynamic resistance, since this affects the potency sensibly consumed by the vehicle. Although the first detailed studies have been initiate in 1920, until today most of the cars possesses a form that takes the waste of potency of the order from 30 to 40%. With relationship to the lateral winds, the project of the drag of the models in production almost disrespects completely them [1]-[5]. These facts are due mainly the following causes:

- Almost completely aerodynamic study should be experimental or numeric, with great expenditure of time and money;
- The best solutions in terms of aerodynamics adapt badly to an automobile project, in design terms and internal space.

The effects of vertical forces, on the vehicle, influence the adherence of each tire and, therefore, the directional behavior of the vehicle under the action of lateral forces as well as the potency that can be transmitted by the wheels and the break capacity. Therefore, your analysis is also very important in the project of vehicles of great acting.

The simulation in wind tunnel is used thoroughly to model real situations of drainages of air. Besides the automotive industry, a great number of applications can be numbered: dispersion of pollutant, studies of pedestrians' comfort and dispersion of particles [6]-[9].

Aerodynamics is the part of the mechanics that studies the movements and interactions of flowing bodies. The study of the drawings of aircrafts, ships, cars, antennas, bridges... kindred of improving your acting's and safety.

II. Materials and Methods

In the wind tunnel, it was accomplished tests with the objective of visualizing the characteristics aerodynamics of two automobiles in different ways. To accomplish that drainage of air a fan that generated a speed exists (measured with anemometer of hot thread) of 4,1m/s and 4,95m/s.

To visualize the path of the air through the cars, in the wind tunnel, smoke was used, obtained with it burns of vegetable oil.

To obtain the smoke vegetable oil is burned at a 20V voltage generated by a 2.5mm wire. The cars were placed inside of the wind tunnel with the drainage of "air-smoke" and photographed, registering like this the path lines around them, in the 3 different speeds.

Although the objective is to do experiments with drainage of air, similar experiments were also accomplished with drainage of water.

To visualize the path of the water through the cars, in the channel of water, a dye was used.

The cars were placed inside of the channel of water with the watercolor drainage and photographed, registering like this the path lines around them. Figs. 1 and 6 show aerodynamic tunnel set up respectively [10]-[16]. It was also projected and built a camera of stabilization with a rectification of flow of the type beehive. They were considered in the dimension of the camera of stabilization the observations of [17], according to which the length of the honeycomb should be from six to eight times the medium diameter of the cell, and the recommendation of [17], that stipulates a strip for the

Elmo Thiago Lins Cöuras Ford is Ph.D. in Mechanical Engineering; M.Sc. in Administration; M.Sc. in Mechanical Engineering; Engineer of Safety; Mechanical Engineer; Automotive Engineer. Universidade Federal do Rio Grande do Norte – UFRN / PPGCEM, Av. Senador Salgado Filho, S/N, 59072-970, Natal - RN, Brazil (e-mail: elmocouras@hotmail.com).

Valentina Alessandra Carvalho do Vale is Ph.D. Student in Electrical Engineering; M.Sc. in Mechanical Engineering; Engineer of Safety; Electrical Engineer. Universidade Federal do Pernambuco – UFPE / PPGEE, Recife – PE, Brazil (e-mail: ale.vale.ssandra@gmail.com).

José Ubiragi Lima Mendes is Ph.D. in Science and Engineering of Materials; M.Sc. in Mechanical Engineering; Mechanical Engineer. Universidade Federal do Rio Grande do Norte – UFRN / PPGEM, Av. Senador Salgado Filho, S/N, 59072-970, Natal - RN, Brazil (e-mail: ubiragi@ct.ufrn.br).

thickness of the wall of the cells between 0,5 and 2,0mm. The beehive was built starting from tubes of PVC that were cut with 200mm of length each one. Those parts were mounted and agglutinated, forming 625 cells of square section 50x50mm. In Fig. 2, the final assembly of the beehive.

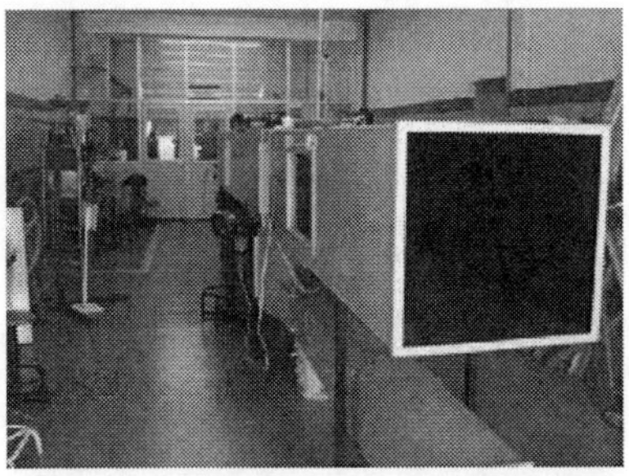

Fig. 1 Aerodynamic Air Tunnel

Fig. 2 Box of harmonization of the type beehive

After the diffuser, the drainage suffers an acceleration, due to the presence of a contraction in which the drainage suffers a pressure loss, that is, turns into kinetic energy (speed) that goes to the rehearsal section. The whole construction of the structures of the tunnel (diffuser, contraction, rehearsal section), they were accomplished by techniques used by [17]. Parts as: The box of harmonization of the drainage; the cooker hood to retain the coming vibrations of the motor and the helix of the motor for generation of the wind, technical were accomplished used at the Laboratory of Aerodynamics. Figs. 3 and 4 show a view of the Pitot tube, the Fig. 5 presents the centrifugal fan used in aerodynamic tunnel.

Fig. 3 Pitot tube

Fig. 4 Height adjustment of the Pitot tube

Fig. 5 Centrifugal fan

Fig. 6 Water of tunnel

Fig. 7 Model 2

Fig. 8 Models 3 and 4

III. Results and Discussion

The presented results were obtained in the experimental rehearsal. In each one of the points, seven measurements were accomplished, of the five were taken advantage of, being eliminated the extreme measures.

In the wind tunnel, it was possible to notice that the drainage of air through the model 1 (Fig. 9) finds smaller resistance than the model 2 (Fig. 10). This happens due to form of the model 1 to be aerodynamic, produced minor I drag and turbulence formation.

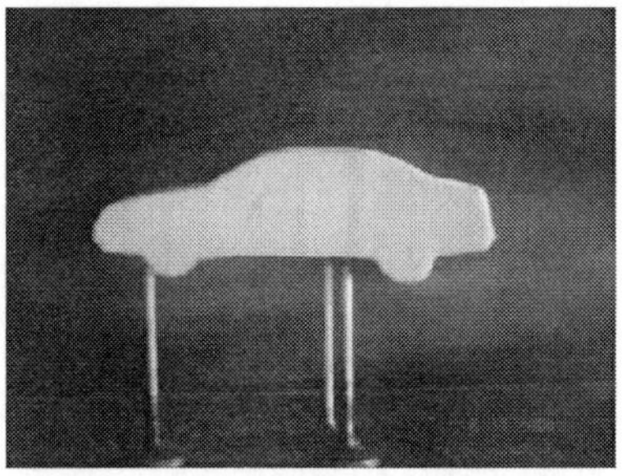

Fig. 9 Text Model 1

Fig. 10 Text Model 2

In the water tunnel, it is observed that the model 4 (Fig. 11) has an upper aerodynamic to model 3 (Fig. 12) due to formation of less turbulence and production of least resistance. In the same way that the model 4 in relation to the model 3, this "better aerodynamics" of the model 4 is due to the fact of possessing a closer form of a "wing" and a smaller front area.

In the automobile branch, the aerodynamics is factor of highest importance that will contemplate later in the acting of the automobile, as economy or stability.

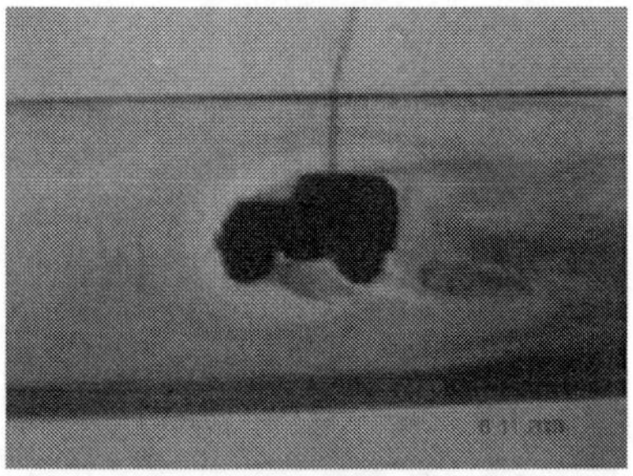

Fig. 11 Text Model 3

Fig. 12 Text Model 4

In a streetcar, the sustentation should not harm the aerodynamic efficiency, for they are obtained larger acting and smaller consumption.

IV. Conclusion

Before the rehearsals accomplished in elapsing of this research, the wind tunnel behaved in a satisfactory way, having reached the objective for which was built, being this of easy construction, simplified operation and of low cost. The expansion camera with beehive was fundamental to minimize the effect of the rotation imposed to the drainage by the axial fan.

Acknowledgment

Thanks go to Coordenação de Pessoal de Nível Superior - CAPES, Conselho Nacional de Desenvolvimento Científico e Tecnológico – CNPQ, Universidade Luterana do Brasil - ULBRA, Universidade Federal de Pernambuco – UFPE and Universidade Federal do Rio Grande do Norte - UFRN, for providing the means to carry out this work.

References

[1] Aerodynamics of road vehicles: from fluid mechanics to vehicle engineering / Syed R. Ahmed; edited by Wolf-Heinrich Hucho. Warrendale: SAE, 1998. 918 p.

[2] Ahmed, S. R. Wake Structure of typical automobile shapes. Journal of Fluids Engineering, Vol. 103, Mar 1981.

[3] Ahmed, S. R. Influence of base slant on the wake structure and drag of road vehicles. Journal of Fluids Engineering, Vol. 105, Dec. 1983. ABNT. 1997. Resíduos Sólidos: ABNT/NBR 10004. Rio de Janeiro.
[4] Anderson, J. D. Fundamentals of Aerodinamics. 3 ed., McGraw-Hill, 2001.
[5] Câmara, E. F. Estudo aerodinâmico do escoamento sobre modelo de ônibus. ITA: São José dos Campos, 1993.
[6] Campbell, C. The sports car. Its design and performance. UK: Chapman and Hall Ltda. 1969.
[7] Canale, A. Automobilística: Dinâmica e Desempenho. Ed. Érika: São Paulo, 1989.
[8] Catalano, F. M. Projeto, construção e calibração em túnel aerodinâmico de circuito aberto tipo N.P.L. de seção transversal hexagonal. USP: São Carlos, 1988.
[9] Carregari, A. L. Estudo do escoamento de ar sobre a carroceria de um ônibus usando programa de CFD e comparação com dados experimentais. USP: São Carlos, 2006.
[10] Diuzet, M. The moving-belt of the I.A.T. Long test section wind tunnel. Journal of Wind Engineering and Industrial Aerodynamics, Vol. 22, pp. 237-244, 1986.
[11] Fox, R.W. Introdução à mecânica dos fluidos. McDonald, LTC: Rio de Janeiro, 2001. 504p.
[12] Hucho, W. H. Aerodynamics of Road Vehicles. Butterworth, 1987.
[13] Kline, S. J.; McClintock, F.A. Describing Uncertainties in single-sample experiments. Mechanical Engineering, Jan. 1953.
[14] Lajos, T.; Preszler, L. Effect of moving ground simulation on the flow past bus models. Journal of Wind Engineering and Industrial Aerodynamics, Vol. 22, pp. 271-277, 1986.
[15] Nicolazzi, L. C; Rosa, E. Introdução à modelagem de veículos automotores. GRANTE-UFSC: Florianópolis, 2001.
[16] Videira, L. C. P. Estudo Experimental de um modelo de ônibus com ênfase no escoamento da esteira. ITA: São José dos Campos, 2001.
[17] Cöuras Ford, E. T. L. Análise quali-quantitativa do deslocamento da camada limite em carrocerias veiculares. UFRN, 2008.

Studies on Race Car Aerodynamics at Wing in Ground Effect

Dharni Vasudhevan Venkatesan, Shanjay K. E., Sujith Kumar H., Abhilash N. A., Aswin Ram D., V. R. Sanal Kumar

Abstract—Numerical studies on race car aerodynamics at wing in ground effect have been carried out using a steady 3d, double precision, pressure-based, and standard k-epsilon turbulence model. Through various parametric analytical studies we have observed that at a particular speed and ground clearance of the wings a favorable negative lift was found high at a particular angle of attack for all the physical models considered in this paper. The fact is that if the ground clearance height to chord length (h/c) is too small, the developing boundary layers from either side (the ground and the lower surface of the wing) can interact, leading to an altered variation of the aerodynamic characteristics at wing in ground effect. Therefore a suitable ground clearance must be predicted throughout the racing for a better performance of the race car, which obviously depends upon the coupled effects of the topography, wing orientation with respect to the ground, the incoming flow features and/or the race car speed. We have concluded that for the design of high performance and high speed race cars the adjustable wings capable to alter the ground clearance and the angles of attack is the best design option for any race car for racing safely with variable speeds.

Keywords—External aerodynamics, External Flow Choking, Race car aerodynamics, Wing in Ground Effect.

I. Introduction

THE motion of air around a moving vehicle affects all of its components in one form or another. Of late, race car aerodynamics gained increased attention, mainly due to the utilization of the negative lift (down-force) principle, yielding several important performance improvements. In all forms of racing, however, aerodynamics eventually surfaced as a significant design parameter. Although the foundations of aerodynamics were formulated over the years, not all principles were immediately utilized for race car design [1]-[4]. In an attempt to resolve some of the external aerodynamics problems of race car and in the light of new findings of wing in ground effect by Sanal Kumar et al. [5], a substantial revision of the existing idea is required for the design of high performance and high speed race cars. Wings in ground effect possess many aerodynamic features of both practical and fundamental importance. In general, the lift and drag forces of a wing will considerably change near the ground. When an airfoil moves near the ground, flow around the airfoil is viscous and has many viscous interactions with the ground. In the analysis of ground effect on the aerodynamic properties of the airfoils, the boundary layer on the airfoil must be considered. On the other hand, prediction of location of the onset of the transition phenomenon, as an example of boundary layer characteristics, is also necessary in order to predict the drag because the skin friction related to a laminar boundary layer is lower than that of a turbulent one [1].

Von Doenhoff [4] investigated boundary layer around the symmetrical NACA airfoil in zero lift condition. His surveys were made considering different Reynolds number based on airfoil's chord. In that work, drag of airfoil, distribution of skin friction over the surface of the airfoil and onset of the transition point were found. Lian and Shyy [2] investigated performance of arigid airfoil and a flexible airfoil by numerical method. All the cited works were focused on the investigations of airfoils at distance far from the ground surface, but limited work has been performed or reported for the airfoil that moves near the ground.

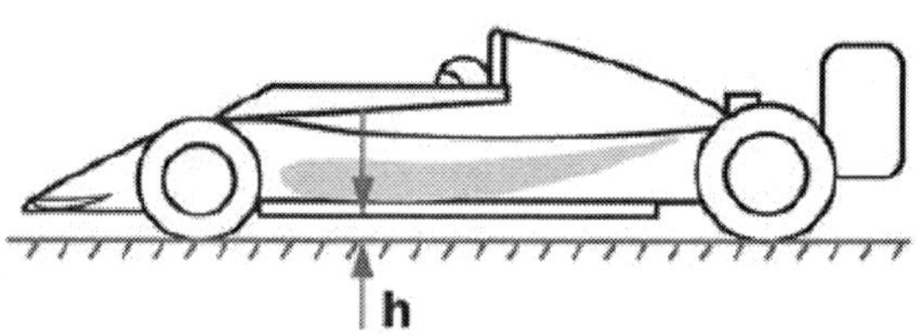

Fig. 1 Race car wings at close proximity of the ground

Wing in ground (WIG) effects is an important phenomenon as it affects the performance of race cars. Fig 1 shows a typical race car with front and rear wings close to ground proximity. Literature review reveals that studies on ground effect on the aerodynamics of a lifting body can be dated as far back as the 1920s but the propulsion aspect of wings in ground effect not yet reported in any open literature. It is generally believed that the presence of ground effect will enhance lift (L) and reduce the drag (D) acting on a lifting body (for example, a wing). However, it is not always true in the light of the findings of Sanal Kumar V. R. et al. [5]. Note that there are a lot of detailed questions which include the details of the relevant flow physics that have yet to be clearly addressed when a race car wings are fixed close proximity to the ground (see Fig. 1). Literature review further reveals that Wing-in-ground effect vehicles operate close to the ground surface by utilizing the air cushion of relatively highly pressurized air created between

Dharni Vasudhevan, Shanjay, Sujith Kumar, Abhilash, and Aswin Ram are Undergraduate Students of the department of Mechanical Engineering, Kumaraguru College of Technology, Coimbatore – 641 049, Tamil Nadu, India (Phone: +91-9894467086, sudhan_619@hotmail.com, keshanjay@ hotmail.com, sujith.kumar093@gmail.com, abhilashajjan@rediffmail.com, ashwin.ram7@ gmail.com).

Sanal Kumar is Professor and Aerospace Scientist and currently with the department of Aeronautical Engineering, Kumaraguru College of Technology, Coimbatore – 641 049, Tamil Nadu, India; (Corresponding Author, phone: +91 – 938 867 9565; + 91 – 875 420 0501, vr_sanalkumar@yahoo.co.in).

International Science Index, Mechanical and Mechatronics Engineering Vol:8, No:7, 2014 waset.org/Publication/9998676

the airfoil and the surface. The aircushion augments lift and reduces drag considerably compared to an out-of-ground effect vehicle; these phenomena further enhance the lift-to-drag ratio, and have been observed during takeoff and landing of aircraft by a number of researchers since the1900s [6-14]. We believe that this may not be true in close proximity of the race car wings with the ground at high speed and certain angles of attack owing to the fact that there is a possibility of change of flow physics due to convergent-divergent channel effect. Earlier studies have shown that the ground has a significant influence on the pressure distributions along the wing surface. As a WIG effect vehicle moves forward, the speed of the oncoming air gradually decreases under the lower wing surface, and dynamic pressure changes to static pressure. This increased pressure is called an air cushion or a ram effect. Sanal Kumar et al. [5] observed that there are possibilities of external flow choking while aircraft flies in close proximity of the ground. The fact is that if the ground clearance height to the chord length of the aircraft is too small, the developing boundary layers from either side of the surface (ground and wing) can interact and develop a transient *fluid-throat*, leading to a choked flow. The results of this study lead to say that one should ensure a safe combination of the dominating parameters for the external flow choking such as vehicle speed, topography and ground clearance, angle of attack, and incoming flow features and coupled geometry of ground and the wing leading to catastrophic failures of race car wing in ground effect. This study is a pointer to the designers of high speed race car with low wings for the safe racing. Note that in most forms of racing it is desirable to create the fastest vehicle in a particular category. In the light of the aforesaid findings one can comprehend that aircushion or ram effect is only in certain combination of the parameters such as ground clearance (h/c), speed and angle of attack.

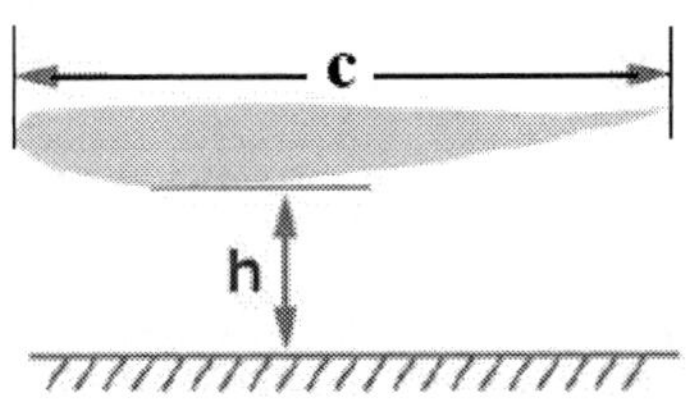

Fig. 2 Physical model of a race car wing

Traditionally, the effects of external aerodynamics are summarized in terms of drag, lift, and stability. Usually the side force (due to aerodynamic side slip) was not examined carefully because race cars go much faster than the prevailing winds, and, instead of lift, the generation of efficient down force became the main issue. The three aerodynamic moments came to light when designers realized that vehicle stability (and handling) can be improved by properly balancing the down force (e.g., front/rear) on the tires. Such desirable aerodynamic down force can be generated by adding lifting surfaces onto, or by modifying, the vehicle's body. When a vehicle moves fast, lateral instability may become uncomfortable from the driver's point of view.

The objective of the present investigation is to carry out parametric analytical studies to better understand the flow physics of WIG effects of race car at different ground clearance, speed and the angles of attack of the race car wings. Fig. 2 shows the physical model of a typical race car wing. The ground clearance h/c (where h is the distance between the lifting body and the ground, and c is the chord of the airfoil cross-section of the wing), and the race car speed, baseline values, are selected based on typical values from literature pertaining to a Formula Mazda car [11]. Vignesh et al. [15] reported that every single surface of a modern Formula One car, from the shape of the suspension links to that of the driver's helmet - has its aerodynamic effects considered. The authors further conjectured that the prudent selection and fixing of a flap at the trailing edge of the race car wing with suitable orientation for increasing the downward force is a meaningful objective for redesigning the race cars for high performance applications. Authors concluded that the maximum design speed of any race car can be increased further by 20 % without any lift penalty by fixing a gurney flap, of size 1% of chord length, fixed vertical to the chord at the trailing edge of the rear wing. Although this paper is of topical interest the authors did not consider the wing in ground effect phenomena. However in this connected paper detailed numerical studies on race car aerodynamics at wing in ground effects have been carried out using the 3d standard k-epsilon turbulence model.

II. Numerical Method of Solution

Numerical simulations have been carried out with the help of a steady 3d, double precision, pressure-based, standard k-epsilon turbulence model. This model uses a control-volume based technique to convert the governing equations to algebraic equations. The viscosity is determined from the Sutherland formula. The airfoil geometric variables and material properties are known *a priori*. Initial wall temperature and inlet temperature are specified. At the exit, far field boundary condition is prescribed. At the solid walls no-slip boundary condition is imposed. The Courant-Friedrichs-Lewy number is chosen as 5.0 in all of the computations. The turbulent kinetic energy and the specific dissipation rate are taken as 0.8. Ideal gas was selected as the working fluid.

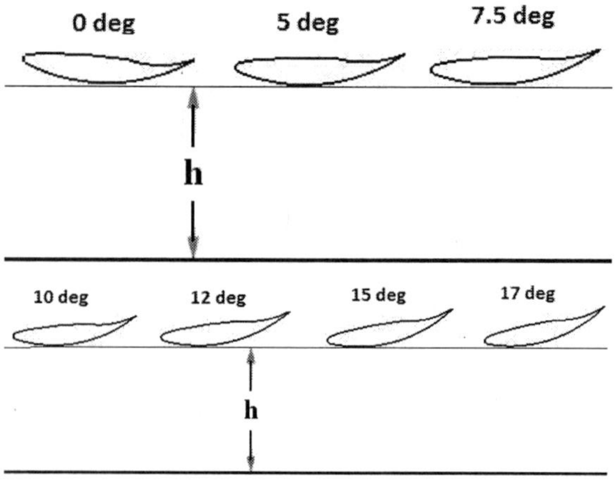

Fig. 3 Race car wing at different angles of attack

Fig. 3 shows the cross sectional view of the race car wing at different angles of attack. Figs. 4 (a), (b) show the typical grid system in the computational domain. The grids are clustered near the solid walls using suitable stretching functions for capturing the boundary layer effect.

(a) Grid system in 3d computational domain

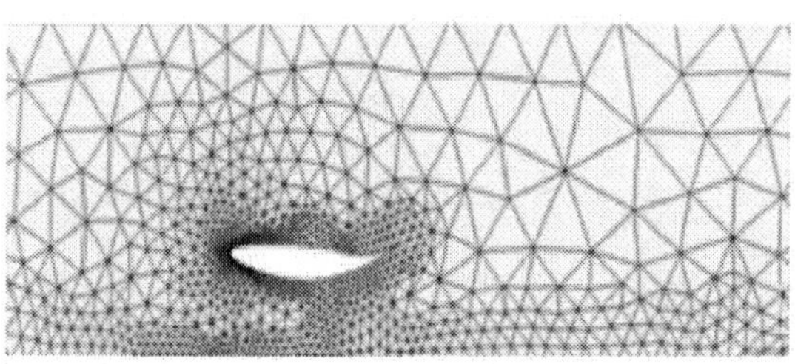

(b) Grid system in 2d plane

Fig. 4 (a), (b) Grid system in the computational domain

III. Results and Discussion

In the parametric analytical studies a typical airfoil has been used to understand the propulsion and the aerodynamics aspects of wing in ground effects (ground effects or surface effects). It occurs typically when the wing of an aircraft flies in close proximity of the ground (approximately one quarter the height of the wing span depending on the aspect ratio of the wing). There are a number of changes in the flow characteristics around the wing due to this effect. In the light of the non-cataloged failures of aircraft and race cars in the ground effect region; detailed numerical studies have been carried out to examine the flow features at wing in ground effect of a typical race car. In this paper using a steady three-dimensional k–epsilon turbulence model, detailed parametric studies have been carried out to examine the external flow features of a race car wing with different angles of attack, ground clearance (h/c), and speed. In the numerical study, a fully implicit finite volume scheme of the compressible, Reynolds-averaged, Navier–Stokes equations is employed. Through detailed parametric analytical studies we have observed different aerodynamic flow features of the race car wings in close proximity of the ground (see Figs. 5-9).

We have conjectured that at a particular speed and ground clearance (h/c) of the wing, the downward force (negative lift) was found high at a particular angle of attack. Note that the flow development and the boundary layer thickness over the lower surface of the wing and the ground are the unique features of each race car at different speed conditions and topography. These are discernible in Figs. 5, 6. Having known this flow physics the next step will be to design a race car wing with flexible and/or adjustable wings capable to alter h/c and the angles of attack for getting the high performance at variable speeds. We have comprehended that any inappropriate combination of speed, ground clearance and angle of attack the possibility of altering the benign aerodynamics characteristics of a race car including the lift (upward force) will be high (see Figs. 7-9). As stated in the introduction external aerodynamic choking (i.e., flow choking between the ground and the wing) will create supersonic flow due to the formation of a transient convergent – divergent (CD) nozzle effect, which can lead to the formation of shock waves. The fact is that if the ground clearance height to the chord length of the aircraft is too small, the developing boundary layers from either side of the surface (ground and wing) can interact and develop a transient *fluid-throat*, leading to a choked flow. Note that due to the choking phenomena the flow passing over the lower surface of the wing will be get accelerated due to the transient formation of a CD nozzle shaped geometry effect between the ground and the lower surface of the wing and consequently a sudden drop in downward force will be the consequent effect. This is an area that needs to be examined carefully, which however not attempted in this paper.

(a) h/c = 0.26

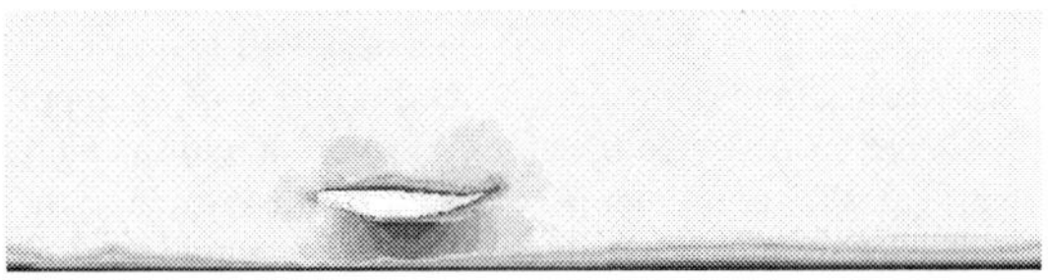

(b) h/c = 0.29

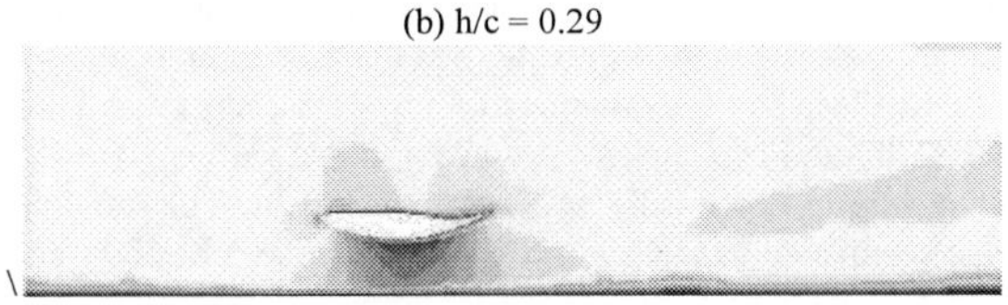

(c) h/c = 0.31

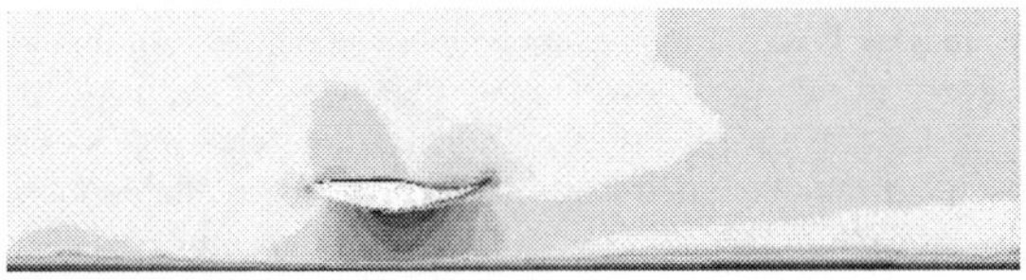

(d) h/c = 0.34

(e) h/c = 0.36

Fig. 5 (a)-(e) Velocity contours at different ground clearance (h/c) at zero angle of attack

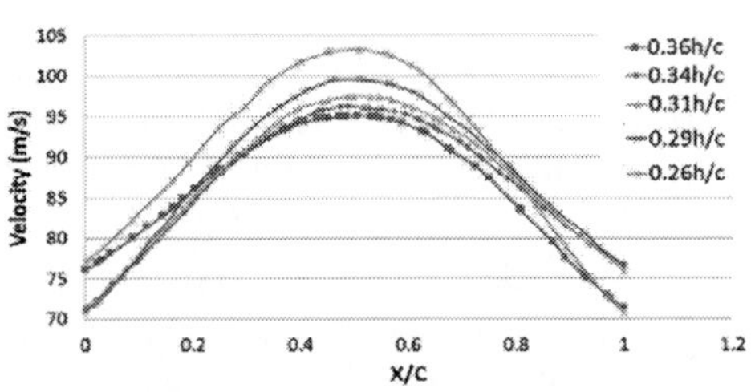

(a) Velocity variations at zero angle of attack

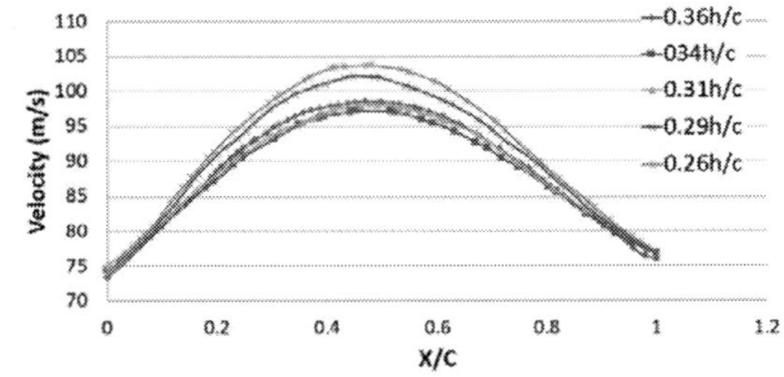

(b) Velocity variations at 5^0 angle of attack

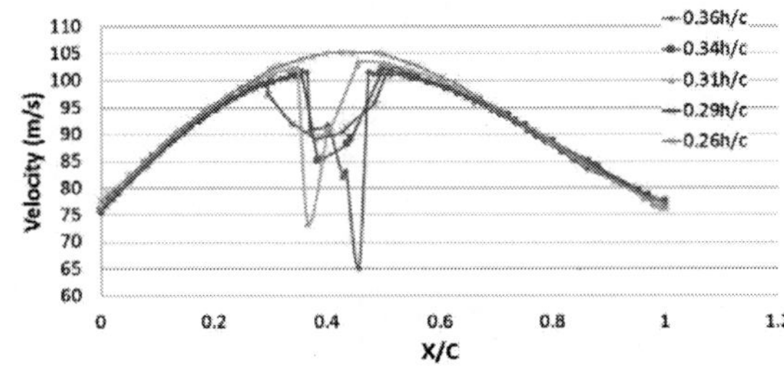

(c) Velocity variations at 7.5^0 angle of attack

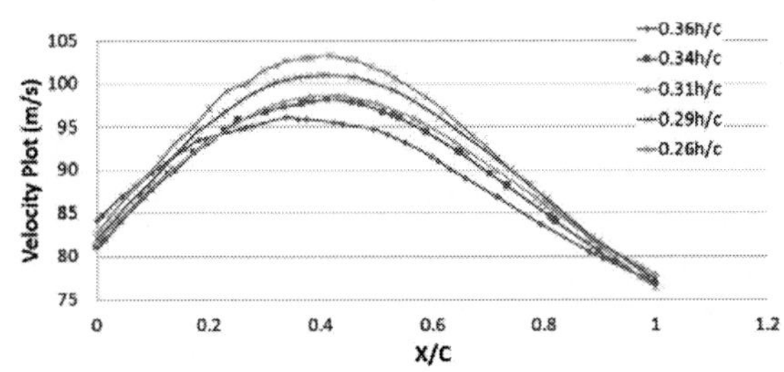

(d) Velocity variations at 10^0 angle of attack

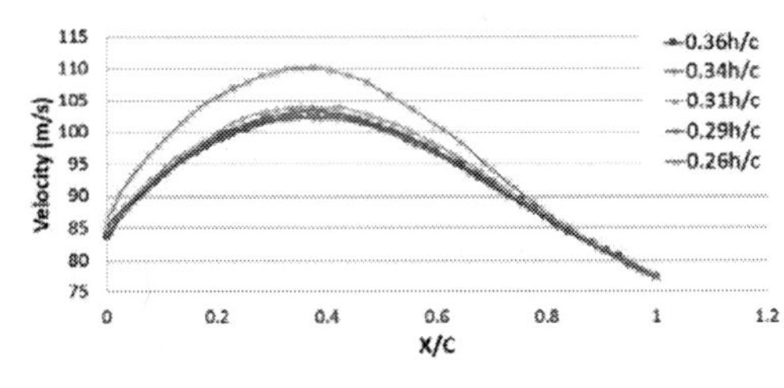

(e) Velocity variations at 12.5^0 angle of attack

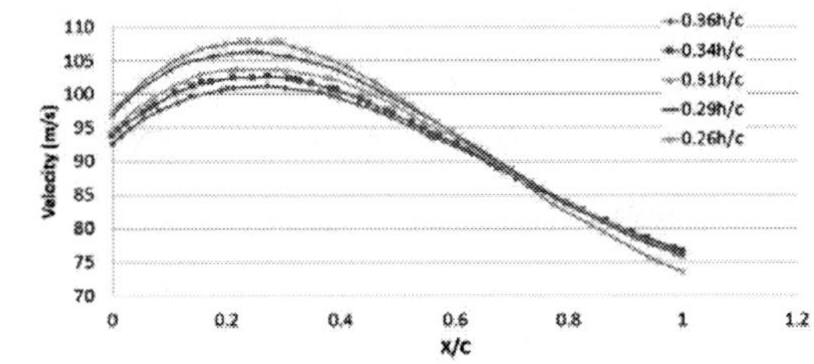

(f) Velocity variations at 15^0 angle of attack

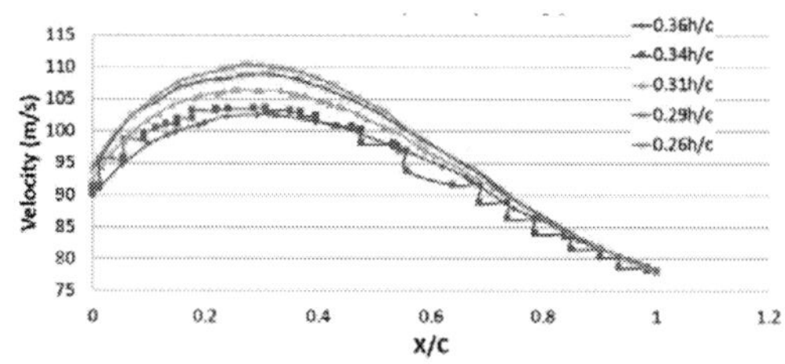

(g) Velocity variations at 17^0 angle of attack

Fig. 6 (a)-(g) Comparison of the velocity variations at different ground clearance at different angles of attack

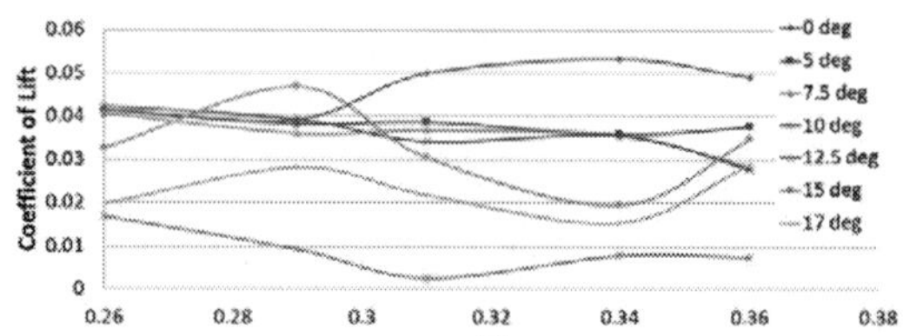

Fig. 7 Comparison of the lift coefficients at different angles of attack and ground clearance

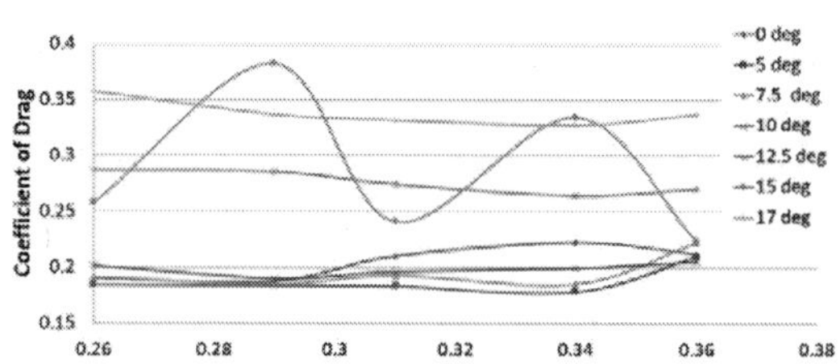

Fig. 8 Comparison of the drag coefficients at different angles of attack and ground clearance

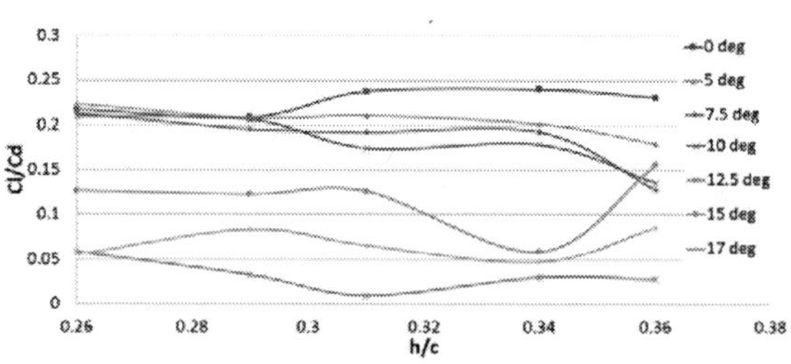

Fig. 9 Comparison of the ratio of lift to drag coefficients at different angles of attack and ground clearance

Furthermore, one should ensure a safe combination of the dominating parameters for the external flow choking such as vehicle speed, topography and ground clearance, angle of attack, and incoming flow features and coupled geometry of ground and the wing leading to possible catastrophic failures of race car wing in ground effect.

It may be recalled here that, owing to the viscous friction, a boundary layer could be formed on the lower walls of the wing and the ground, and their thickness will increase in the downstream direction leading to the formation of a temporary CD nozzle throat at the transition location due to the coupled effects of the port geometry and the area fraction being blocked by the boundary-layer displacement thickness. As a result, in general, at larger distances from the inlet section of the wing the velocity will be high at the transition location and this will lead to high downward force. However, it can lead to the formation of shock waves in the downward region of the wing region having divergent channel effect. This obviously will contribute to the high-pressure spike, which might cause structural failures of race car wing, in addition to the downward force loss. The fact is that if ground clearance height to chord length (h/c) is too small, the developing boundary layers from either side of the port can interact, leading to a choked flow. On the other hand, if the developing boundary layers are far enough apart, then choking does not occur. It may be noted here that most of the available models do not capture the shock wave phenomena encountered in race car at wing in ground effect. Nevertheless, the accurate evaluation of the Mach number contours between the wing and the ground are sufficient to propose the possible occurrence of shock waves at wing in ground effect, which is beyond the scope of this paper. From the point of view of continuum theory, shocks can be treated as localized discontinuities within the flow, which everywhere else satisfies the continuum hypothesis. The authors have inferred that the shock waves at wing in ground effect can generate additional turbulence. The shock waves and the new turbulence level will alter the race car aerodynamics at the wing in ground effect, which merit further investigation.

IV. Concluding Remarks

We have concluded that when the high speed race car wings are at the close proximity of the ground at a particular speed and angle of attack the aerodynamic performance characteristics of the race car, essentially the lower surface, can alter leading to a catastrophic failure. Therefore a safe ground clearance must be ensured throughout the racing, which obviously depends upon the coupled effects of the topography, wing orientation with respect to the ground, the incoming flow features and/or the race car speed. Through these parametric analytical studies we have concluded that for the design of high performance and high speed race cars adjustable wings capable to alter the wing ground clearance (h/c) and the angle of attack is the best design option for racing safely with variable speeds.

Acknowledgment

The authors would like to thank Joint Correspondent, Shankar Vanavarayar of Kumaraguru College of Technology, Coimbatore, India for his extensive support of this research work.

REFERENCES

[1] A. Silverstein, J. Becker, "Determination of boundary layer transition on three symmetrical airfoils in the NACA full-scale wind tunnel", TR No. 637, NACA, 1939.

[2] Y. Lian, W. Shyy, "Laminar–turbulent transition of a low Reynolds number rigid or flexible airfoil", AIAA Journal 45 (7) (July 2007) 1501–1513.

[3] Y. Takahashi, M. Kikuchi, K. Hirano, "Analysis of ground effect on aerodynamic characteristics of aerofoils using boundary layer approximation", JSME International Journal, Series B 49 (2) (2006) 401–409.

[4] A.E. Von Doenhoff, "Investigation of the boundary layer about a symmetrical airfoil in a wind tunnel of low turbulence", Langley Memorial Aeronautical Laboratory, W.R. No. L.507, NACA, 1940.

[5] V.R.Sanal Kumar, Ganesh Shankar, Saravanan.A, SanthoshKumar.N, Lenin. Cand Sathyan.P, "Phenomenological Introduction of Propulsion Aspect of External Flow Choking at Wing in Ground Effect," 50th AIAA Joint Propulsion Conference, Ohio, USA, July 2014. (To appear)

[6] K.Kikuchi.,F.Motoe, M.Yanagizawa, 1997. "Numerical simulation of the ground effect using the boundary element method". Int. Journal for Numerical Methods in Fluids 25, 1043–1056.

[7] H.J.Kim, H.H.Chun, 1998."Designof2-dimensional WIG section by an on linear optimizationmethod".J.Soc.NavalArchit.Korea35 (3), 50–59.

[8] Y.J. Kim, C.Y.Joh, 2004."Aerodynamic design optimization of air foils for WIG vehicle using response surface method". J. Korean Soc. Aeronaut. Space Sci.33 (5), 18–27.

[9] C.Y. Joh, Y.J.Kim,2004."Computational aerodynamic analysis of airfoils for WIG (airfoil-in-ground-effect)-craft". J. Korean Soc. Aeronaut. Space Sci.32 (8), 37–46.

[10] S.C. Luo, Y.S. Chen, "Ground effect on flow past a wing with a NACA0015 cross-section", Experimental Thermal and Fluid Science 40 (2012) 18–28.

[11] W. Kieffer, S. Moujaes,, N. Armbya, "CFD study of section characteristics of Formula Mazda race car Wings" Mathematical and Computer Modeling, 43, 2006, 1275-1287.

[12] S.C. Luo, and Y.S. Chen, "Ground effect on flow past a wing with a NACA0015 cross-section," Experimental Thermal and Fluid Science 40 (2012) 18–28.

[13] J. Wu, and N. Zhao, "Ground Effect on Flapping Wing", 7th Asian-Pacific Conference on Aerospace Technology and Science, 7th APCATS 2013, Procedia Engineering 67 (2013) 295 – 302.

[14] M.H. Djavareshkian, A. Esmaeli, "Smart Spoiler for Race Car", World Academy of Science, Engineering and TechnologyVol: 5 2011-01-23.

[15] J. Vignesh, K.Krishnaraj. G.Sandeep. K.Sabarinath.K,, Merlin Baby Jose, V.R.Sanal Kumar, "Numerical Studies on Aerodynamics of High Performance Race Car," World Academy of Science, Engineering and Technology: Proceedings of the International Conference on Production and Manufacturing Engineering, London, United Kingdom, May, 26-27, 2014.

3D Numerical Studies on External Aerodynamics of a Flying Car

Sasitharan Ambicapathy, J. Vignesh, P. Sivaraj, Godfrey Derek Sams, K. Sabarinath, V. R. Sanal Kumar

***Abstract*—**The external flow simulation of a flying car at take off phase is a daunting task owing to the fact that the prediction of the transient unsteady flow features during its deployment phase is very complex. In this paper 3D numerical simulations of external flow of Ferrari F430 proposed flying car with different NACA 9618 rectangular wings have been carried. Additionally, the aerodynamics characteristics have been generated for optimizing its geometry for achieving the minimum take off velocity with better overall performance in both road and air. The three-dimensional standard k-omega turbulence model has been used for capturing the intrinsic flow physics during the take off phase. In the numerical study, a fully implicit finite volume scheme of the compressible, Reynolds-Averaged, Navier-Stokes equations is employed. Through the detailed parametric analytical studies we have conjectured that Ferrari F430 flying car facilitated with high wings having three different deployment histories during the take off phase is the best choice for accomplishing its better performance for the commercial applications.

***Keywords*—**Aerodynamics of flying car, air taxi, negative lift, roadable airplane.

I. Introduction

THE modern flying car concepts like the Terrafugia Transition are showing remarkable promise. The Terrafugia Transition is a light sport, roadable airplane under development by Terrafugia since 2006 [1]-[3]. Although flying car concept has been around since the early days of motoring, the optimized aerodynamic design of a flying car is still a daunting task [1]-[15]. Of late the concept of Godfrey. D. S. et al. [1] on the design of Ferrari F430 flying car got reasonable attention in the industry. The authors reported that Ferrari F430 flying car with thrust to weight ratio 0.3176 will take off at 53 km/hr with the help of NACA 9618 airfoil shaped wings deployed in both sides. It is well known that the existing flying car designs are expensive and additionally the previous studies reveal that more efforts must be put for the realization of a commercial flying car. The literature review reveals that the designers have been trying to build a flying car for a century, but only a few designs ever succeeded in flying through the air and driving on the road. The literature review further reveals that more efforts must be put for the realization of a lucrative design of a flying car. Since the integration of computational fluid dynamics (CFD) methods into a wide range of engineering disciplines is rising sharply, mainly due to the positive trends in computational power and affordability, one can use it in the flying car design lucratively. Note that the aerodynamics design of a flying car is more complex than high performance race car and a conventional aircraft owing to the fact that prediction of the transient unsteady flow features during the deployment of flying car wings and further its take off phase is very complex.

The designer of a flying car has the major concern on the creation of upward force (lift) with low takeoff velocity, with minimum drag, coupled with better stability and control warranting its overall high performance both in road and air. Note that the amount of upward force created by the flying car is depending primarily on two things; viz., the shape, including surface area, aspect ratio and cross-section of the device, and the device's orientation (or angle of attack). This paper is a continuation of the work of Godfrey. D. S. et al. [1]. In this connected paper authors made several attempts to make Ferrari F430 car to fly with lower take-off velocity by using cascading of wings for escaping from the occasional traffic congestions. Fig. 1 shows the physical model and the Mach number contours of the Ferrari F430 flying car proposed by Godfrey. D. S. et al., [1]. The authors successfully carried out 3D CFD analyses using a k-omega turbulence model. Although CFD is now widely used in automobile and aerospace industry, archival reports on flying cars are less than abundant. In this paper 3D numerical studies on external flow features of a Ferrari F430 car with different wing positions and deployment history are reported. Note that validated CFD results can be used as a preliminary design tool or to complement experimental methods.

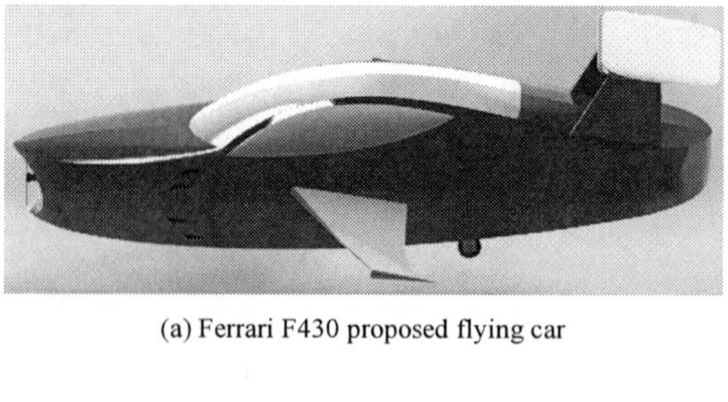

(a) Ferrari F430 proposed flying car

Sasitharan Ambicapathy, Vignesh J, Sivaraj P, and Godfrey Derek Sams are with Kumaraguru College of Technology, Coimbatore – 641 049, Tamil Nadu, India, (e-mail: ambi.sasitharan@gmail.com, vigneshj91@gmail.com, ariensiva@gmail.com, godfreyrocz@gmail.com)

Sabarinath K was with Kumaraguru College of Technology, India, and currently with Department of Aerospace Engineering, University of Cranfield, UK, (e-mail: s.karunakaran@cranfield.ac.uk)

V. R. Sanal Kumar is currently with Department of Aeronautical Engineering, Kumaraguru College of Technology, Coimbatore – 641 049, Tamil Nadu, India; (Corresponding Author, phone: +91 – 938 867 9565; fax: +91-422-266 9406; email: vr_sanalkumar@yahoo.co.in).

(b) Mach number contours of Ferrari F430 flying car

Fig. 1 (a), (b) Demonstration of the proposed Ferrari F430 flying car (Adopted from Godfrey et al. [1])

II. Numerical Method of Solution

Numerical simulations have been carried out with the help of a three-dimensional standard k-omega turbulence model. This turbulence model is an empirical model based on model transport equations for the turbulence kinetic energy and a specific dissipation rate. This code solves standard k-omega turbulence equations with shear flow corrections using a coupled second order implicit unsteady formulation. In the numerical study, a fully implicit finite volume scheme of the compressible, Reynolds-Averaged, Navier-Stokes equations is employed. Initial wall temperature, inlet total pressure and temperature are specified. Ideal gas is selected as the working fluid. The code has successfully validated with the help of benchmark solutions. The wing is designed with a NACA series of 9618 airfoil due to its better aerodynamic characteristics. The total cell count of Ferrari F430 flying car base model is around 1,60,000.

Fig. 2 shows the idealized 3D symmetrical model (right half) of Ferrari F430 (base model) reference car. Figs. 3-9 show different physical models of Ferrari F430 flying car fixed with different types of wings for examining it external flow features. Figs. 4 (a)-(c) show Ferrari F430 flying car with high wing having three different deployment histories. The base wing can be fixed or retractable. This design was proposed for meeting the road width dimensional constraints. Note that at the time of taxiing, Ferrari F430 is having only base wings with an acceptable width of the road. Further at the time of takeoff first deployment of both side wings will take place and later during the climb and cruise condition second deployment will take place.

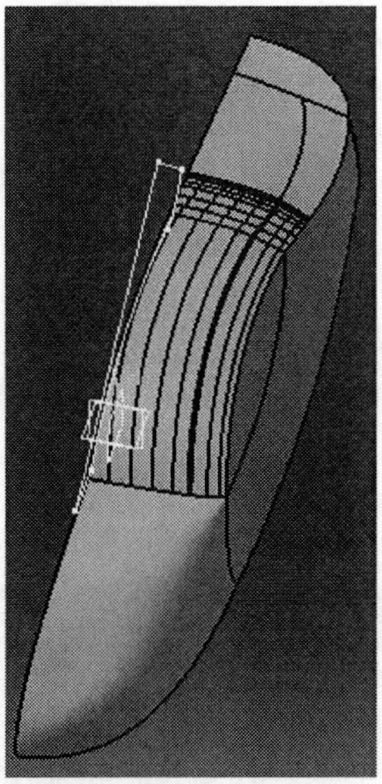

Fig. 2 The 3D base model (right half) of Ferrari F430

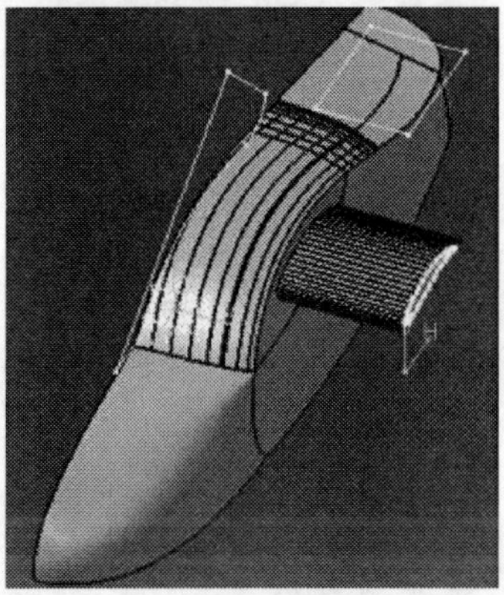

Fig. 3 Ferrari F430 flying car with high wing (Case-1)

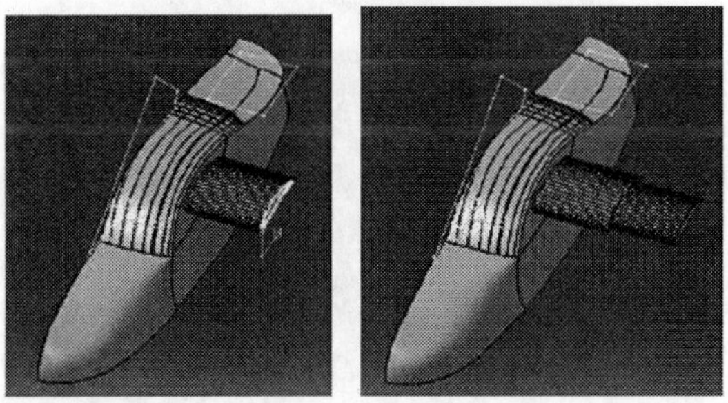

(a) Base wing

(b) Base wing with first extension

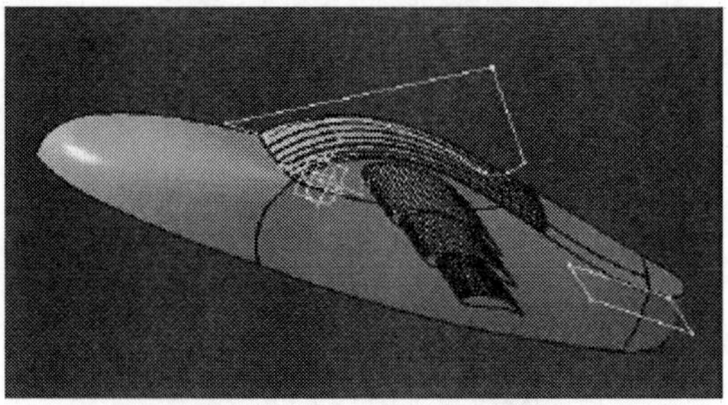

(c) Takeoff position after the second extension of the wing

Fig. 4 (a)-(c) Ferrari F430 flying car with high wing having different deployment histories (Case-2)

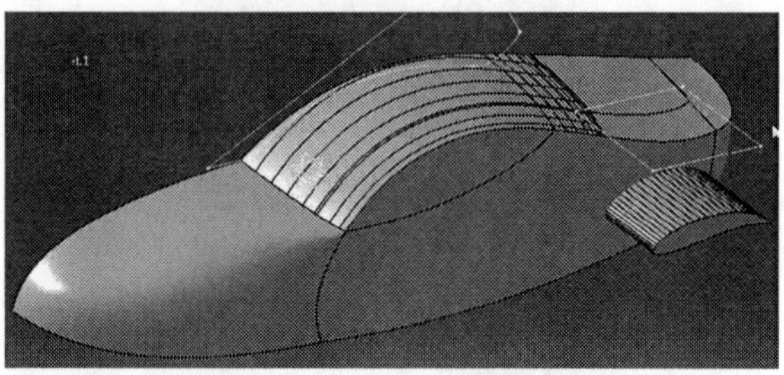

Fig. 5 Ferrari F430 flying car with wing near to the rear end (Case-3)

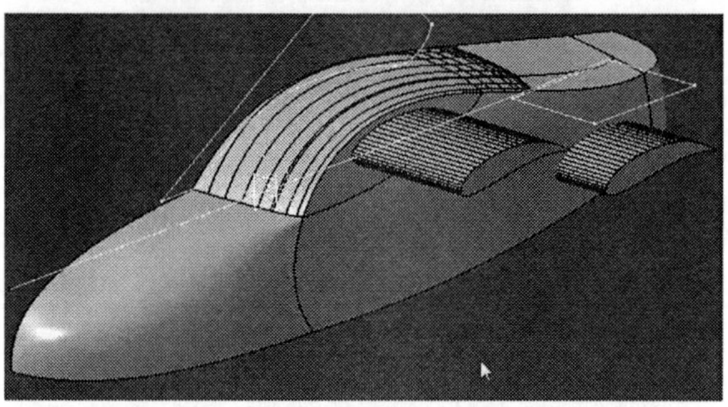

Fig. 6 Ferrari F430 flying car having both high and rear end wings (Case-4)

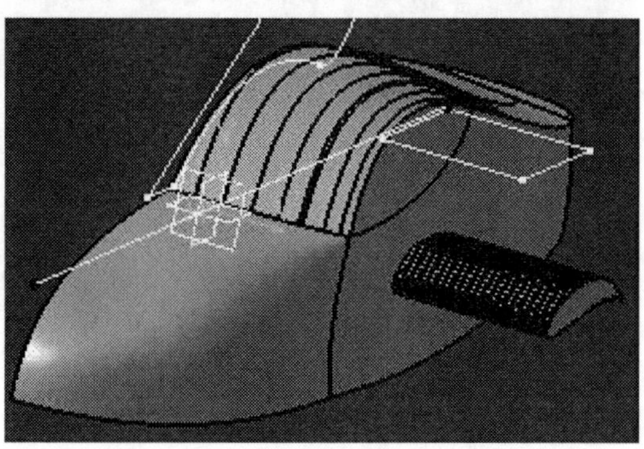

Fig. 7 Ferrari F430 flying car with mid wing (Case-5)

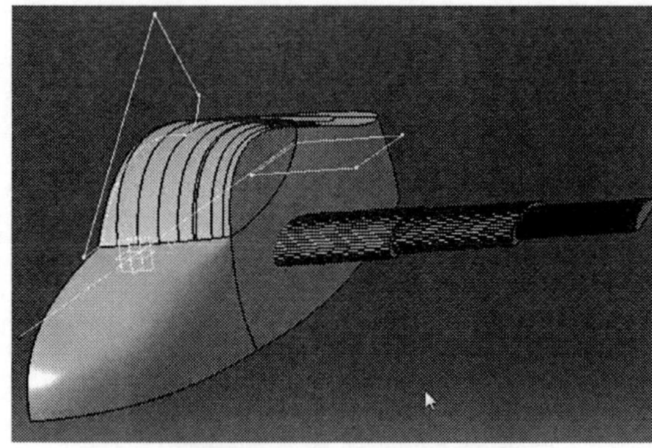

Fig. 8 Ferrari F430 flying car with mid wings having three different deployment histories (Case-6)

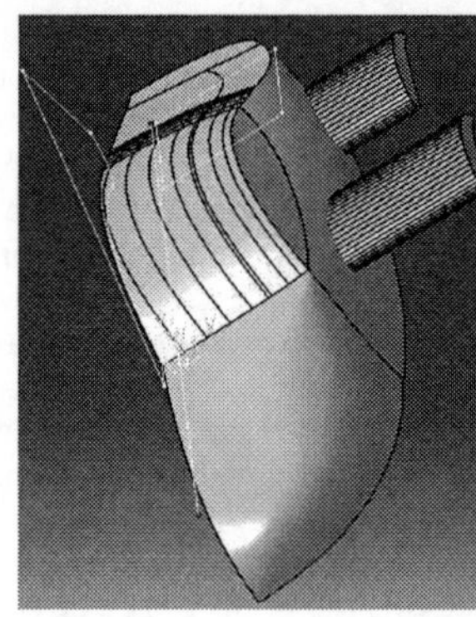

Fig. 9 Ferrari F430 flying car having both mid and rear end wings (Case-7)

III. Results and Discussion

The design of a commercial flying car must be in line with the present road scenario. Therefore, selection of the dimensions of the wing is one of the major constraints of any flying car designer. Admittedly, fixing big wings to the road-able car can easily generate desirable lift at desirable take off velocity for its flying. But optimum design of a flying car for both road taxiing and air taxiing is a daunting task. Due to the dimensional constraints of roads, in this paper authors have made several attempts for creating desirable lift at minimum take off velocity with different wing positions and angles of attack. After detailed parametric analytical studies we observed that a case (Case-2) with deployable high wing, as seen in Figs. 4 (a)-(c), can meet the road dimensional constraints and can produce desirable lift at low take off velocity. In this paper several case studies have been carried out with high wing, mid wing and rear wing for evaluating the take-off velocity of Ferrari F430 car with NACA 9618 rectangular wing. Figs 10 (a)-(c) show the contours of static pressure at various speeds of Ferrari F430 car (right half). Grid for this model has been generated after a detailed grid refinement exercises. The grids are clustered near to the wall and suitable CFL number has been chosen though out the computation. The results are verified through the benchmark solutions. Table I shows the aerodynamics characteristics of Ferrari F430 car without wing. It is evident from Table I that without wing Ferrari F430 cannot take off at a free stream velocity of 45 m/s due to the negative lift coefficient, warranting better road performance.

TABLE I
AERODYNAMIC CHARACTERISTICS OF FERRARI F430 CAR (BASE MODEL)

Velocity	C_L	C_D
15 m/s	-0.32159	0.334296
30 m/s	-0.29422	0.318038
45 m/s	-0.28424	0.312268

C_l – Lift Coefficient C_d – Drag Coefficient

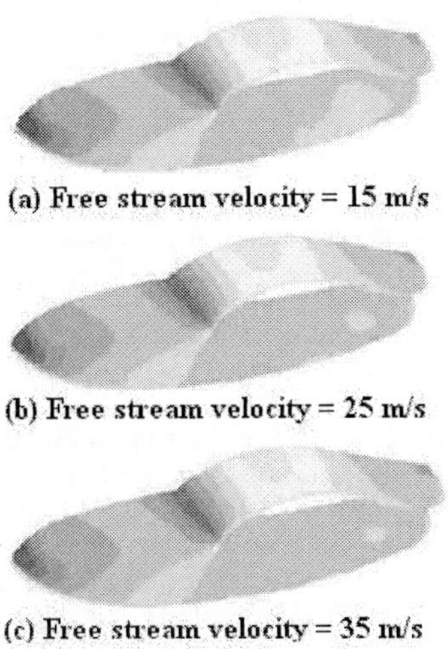

Fig. 10 (a)-(c) Contours of static pressure at various speeds of Ferrari F430 reference car

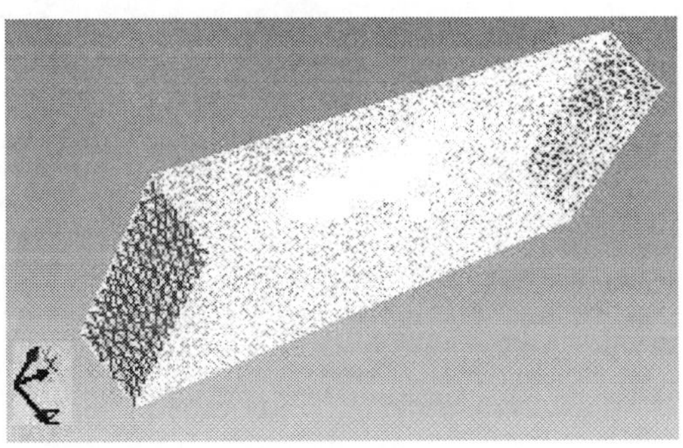

(a) Full computational domain.

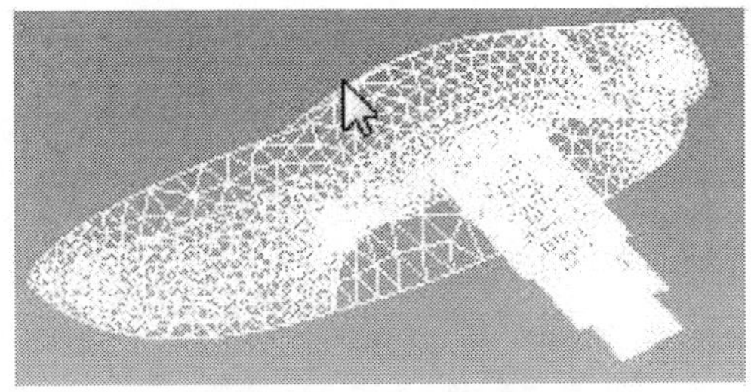

(b) Grid model of Ferrari F430 flying car (Case-2)

Fig. 11 (a-b) Grid system in the computational domain

As stated in the introduction, in this paper the principle that allows a car to rise off the ground by creating lift using additional wings. The primary components of a flying car can be used to create upward force when the car is travelling at take off velocity. This can be achieved by streamlining the car body, and the use of suitable wings. The rounded and tapered shape of the top of the car is designed to slice through the air and minimize wind resistance. In this paper authors have made several attempts to create lift higher than the weight of the car by fixing the rectangular NACA 9816 wing at different positions on the body with different angles of attack. Figs. 11 (a), (b) show the grid system in the computational domain. The numerical results generated for various cases are presented in Figs. 12-16.

Airfoils are recognized as lifting surface objects and a particular criterion called performance parameter is defined for them. This parameter is the ratio of lift-to-drag of the airfoil. Since these objects move in viscous fluid, a particular flow regime with low thickness and larger velocity gradient is formed around them, which is identified as boundary layer. This flow regime causes several changes in behavior of the airfoils such as the inception point of the transition phenomenon, the starting point of flow separation, and the boundary layer thickness. For capturing this intrinsic flow physics the aerodynamic characteristics of the flying car with NACA 9618 wing at different angles of attack at different velocities have been evaluated in viscous flows and reported in Tables II-VIII.

Analyses reveal that at the given range of free steam velocity Ferrari F430 flying car with different geometrical options considered in this study will generate upward force. However analyses further reveal that at these free stream velocities Ferrari F430 car with deployable NACA 9618 rectangular wings (case-2) can generate lift higher than the weight of the flying car at the lower take off velocity, warranting low taxiing distance during traffic congestion.

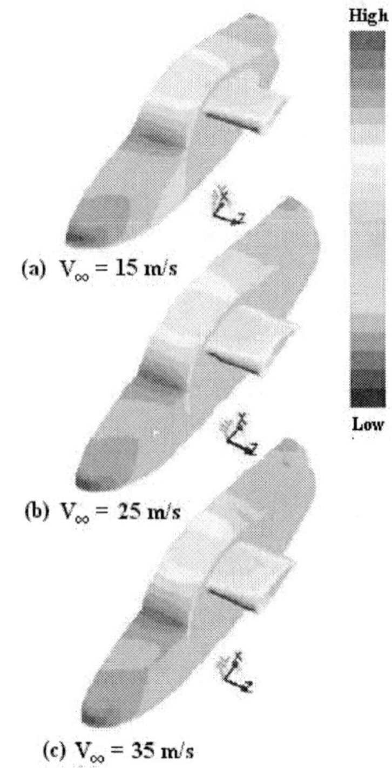

Fig. 12 (a)-(c) Contours of the static pressure at various speeds of Ferrari F430 flying car with short high wing model

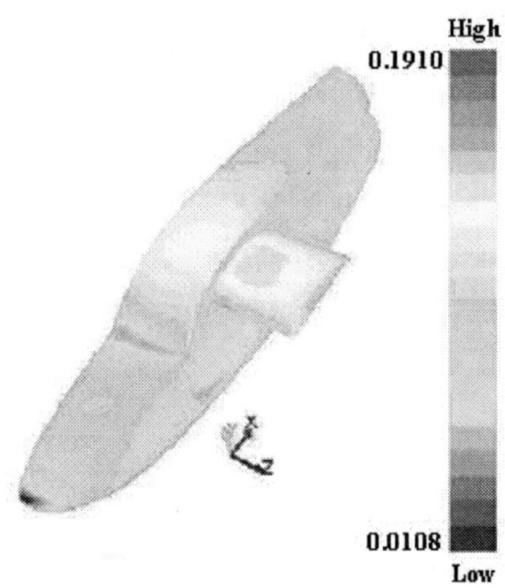

Fig. 13 Contours of Mach number at a free stream velocity of 35 m/s of Ferrari F430 flying car with short high wing model

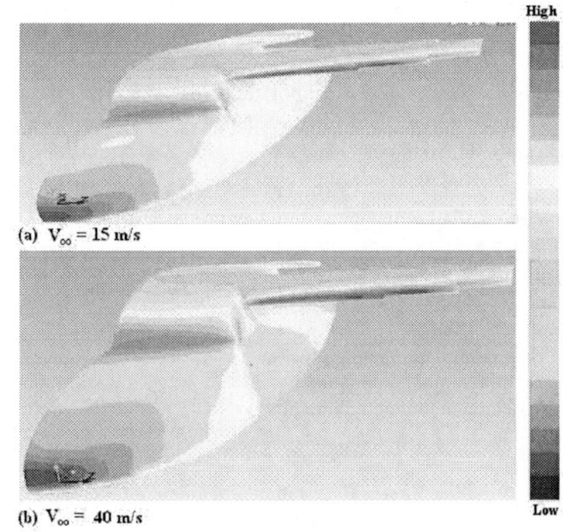

Fig. 14 (a)-(b) Contours of the static pressure at two different speeds of Ferrari F430 flying car with long (extendable) high wings

Table I shows the aerodynamic characteristics of Ferrari F430 car at three different speeds. It is evident from the numerical results generated through parametric analytical studies that up to 45 m/s Ferrari car produce negative lift. But Tables II-VIII show that Ferrari F430 car can be converted in to a flying car well before reaching the free stream velocity of 45 m/s with the corresponding wings. We have observed that among the seven physical models of Ferrari F430 flying car, considered in this study, a case of long high wing with three different deployment histories (Case-2) is the best flying car for commercial applications. Figs. 14 (a), (b) show the contours of static pressure at two different speeds of the aforesaid model of Ferrari F430 flying car with high extendable wings. Figs. 15-17 are corroborating that Case-2 is the best flying car option for getting better aerodynamic efficiency.

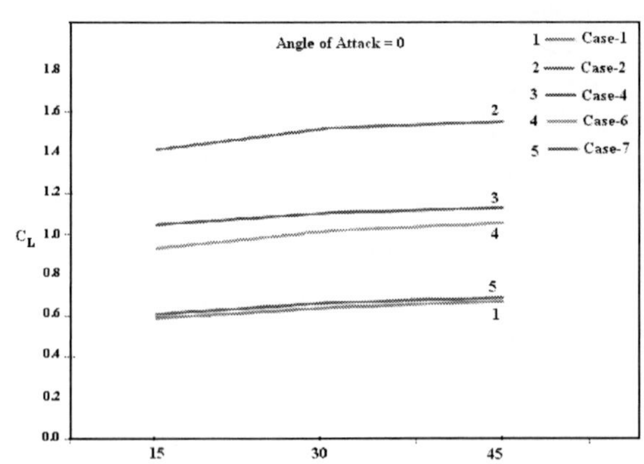

Fig. 15 Comparison of C_L variations for different flying car models at different velocities at zero angle of attack

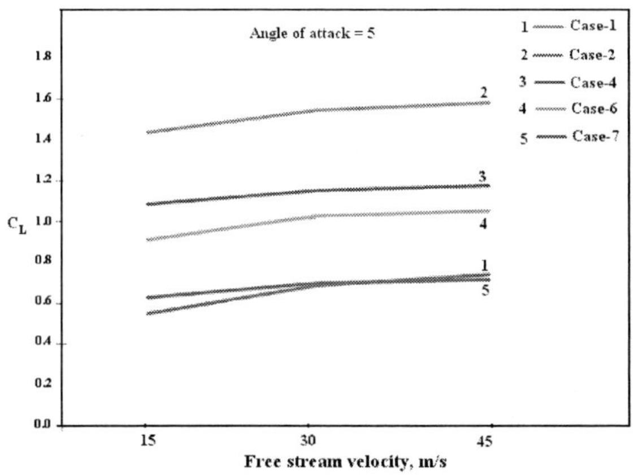

Fig. 16 Comparison of C_L variations for different flying car models at different velocities at an angle of attack of five

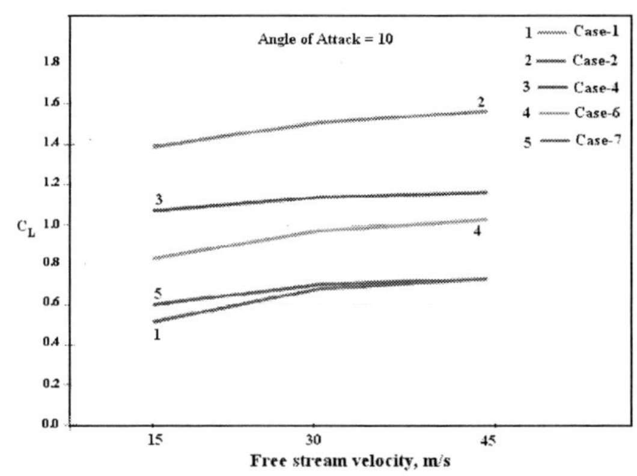

Fig. 17 Comparison of C_L variations for different flying car models at different velocities at an angle of attack of ten

TABLE II
AERODYNAMIC CHARACTERISTICS OF FERRARI F430 FLYING CAR (CASE-1)

V	AoA = 0		AoA = 5		AoA = 10	
	C_L	C_D	C_L	C_D	C_L	C_D
15 m/s	0.591342	0.501152	0.543714	0.503776	0.515916	0.496604
30 m/s	0.643786	0.472916	0.678393	0.485323	0.679274	0.485019
45 m/s	0.673725	0.4583	0.73027	0.483259	0.729097	0.448591

V-Velocity, AoA-Angle of Attack, C_l -Lift Coefficient, C_d-Drag Coefficient.

TABLE III
AERODYNAMIC CHARACTERISTICS OF FERRARI F430 FLYING CAR (CASE-2)

V	AoA = 0		AoA = 5		AoA = 10	
	C_L	C_D	C_L	C_D	C_L	C_D
15 m/s	1.4148661	0.56931906	1.419630	0.573794	1.389823	0.576236
30 m/s	1.5191091	0.53600213	1.525492	0.549478	1.509404	0.542711
45 m/s	1.552219	0.53030393	1.565314	0.530713	1.568809	0.549890

V-Velocity, AoA-Angle of Attack, C_l -Lift Coefficient, C_d-Drag Coefficient.

TABLE IV
AERODYNAMIC CHARACTERISTICS OF FERRARI F430 FLYING CAR (CASE-3)

V	AoA = 0		AoA = 5		AoA = 10	
	C_L	C_D	C_L	C_D	C_L	C_D
15 m/s	0.188866	0.418414	0.194559	0.412738	0.17284	0.398857
30 m/s	0.220425	0.396399	0.267831	0.396057	0.278367	0.40007
45 m/s	0.257569	0.385901	0.287629	0.381799	0.295697	0.370281

V-Velocity, AoA-Angle of Attack, C_l -Lift Coefficient, C_d-Drag Coefficient.

TABLE V
AERODYNAMIC CHARACTERISTICS OF FERRARI F430 FLYING CAR (CASE-4)

V	AoA = 0		AoA = 5		AoA = 10	
	C_L	C_D	C_L	C_D	C_L	C_D
15 m/s	1.049693	0.561152	1.07176	0.562124	1.059179	0.55731
30 m/s	1.106441	0.537064	1.137165	0.547106	1.14011	0.542625
45 m/s	1.13085	0.533458	1.1623	0.529877	1.181761	0.543542

V-Velocity, AoA-Angle of Attack, C_l -Lift Coefficient, C_d-Drag Coefficient.

TABLE VI
AERODYNAMIC CHARACTERISTICS OF FERRARI F430 FLYING CAR (CASE-5)

V	AoA = 0		AoA = 5		AoA = 10	
	C_L	C_D	C_L	C_D	C_L	C_D
15 m/s	0.613128	0.540738	0.620713	0.541608	0.599016	0.538362
30 m/s	0.666163	0.518743	0.691823	0.524213	0.697401	0.521947
45 m/s	0.689381	0.512194	0.707888	0.509476	0.727898	0.518327

V-Velocity, AoA-Angle of Attack, C_l -Lift Coefficient, C_d-Drag Coefficient.

TABLE VII
AERODYNAMIC CHARACTERISTICS OF FERRARI F430 FLYING CAR (CASE-6)

V	AoA = 0		AoA = 5		AoA = 10	
	C_L	C_D	C_L	C_D	C_L	C_D
15 m/s	0.932935	0.546917	0.901326	0.548907	0.833179	0.547809
30 m/s	1.020041	0.515027	1.012353	0.522616	0.968134	0.512653
45 m/s	1.054759	0.50595	1.038988	0.505903	1.028403	0.517948

V-Velocity, AoA-Angle of Attack, C_l -Lift Coefficient, C_d-Drag Coefficient.

TABLE VIII
AERODYNAMIC CHARACTERISTICS OF FERRARI F430 FLYING CAR (CASE-7)

V	AoA = 0		AoA = 5		AoA = 10	
	C_L	C_D	C_L	C_D	C_L	C_D
15 m/s	0.311829	0.4680322	0.314031	0.466983	0.279028	0.46388
30 m/s	0.361834	0.4414857	0.368874	0.447814	0.359085	0.434274
45 m/s	0.381496	0.4332071	0.385492	0.430581	0.401705	0.440808

V-Velocity, AoA-Angle of Attack, C_l -Lift Coefficient, C_d-Drag Coefficient.

TABLE IX
TAKE OFF VELOCITY OF FERRARI F430 FLYING CAR

CASE	TAKE OFF VELOCITY (Km/hr)
Case-1	56
Case -2	36
Case-4	42
Case-6	43
Case-7	53

We have observed that Ferrari F430 model generated negative lift coefficient at given velocities. But the same model with NACA 9618 wing generated positive lift, which was sufficient for its take off from the road at the given velocity. In this paper detailed analyses of external flow features and the corresponding aerodynamics characteristics of various flying car options have been carried out in detail at various flying conditions, orientations and wing positions. Among the cases considered in this study we have observed that a case with long high deployable wing will generate lift higher than the weight of the flying car for its take off at a lower take off velocity. Effort has been taken for estimating the take off velocities of various models of Ferrari F430 flying car. We have observed that lowest take off velocity is obtained as 36 Km/hr for a case with deployable wing (Case-2). Table IX shows the comparison of take off velocities of various successful models of flying car considered in this study.

IV. Concluding Remarks

We observed that the Ferrari F430 flying car with high extendable NACA 9618 rectangular wing will take off at 36 km/hr. Through the detailed parametric analytical studies we have concluded that Ferrari F430 flying car with long high wing facilitating three different deployment histories during the take off phase is the best choice for its commercial applications.

Acknowledgment

The authors would like to thank the management of Kumaraguru College of Technology, India for their extensive support for completing this research work.

References

[1] Godfrey Derek Sams, Kamali Gurunathan, and Prasanth Selvan, V.R. Sanal Kumar "Conceptual Design and Analysis of Ferrari F430 Flying Car", with proceedings of *"International Journal of Research in Engineering and Technology (IJRET)"* Vol. 1, No. 6, 2012 ISSN 2277 – 4378.

[2] Godfrey Derek Sams, Kamali Gurunathan, and Prasanth Selvan, V.R. Sanal Kumar "Conceptual Design and Analysis of Ferrari F430 Flying Car", International Conference on Aerospace, Automotive and Mechanical Engineering (ICAAME'2012), Planetary Scientific and Research Center, Pattaya, Thailand, December 8-9, 2012.

[3] Page, Lewis, "Terrafugia flying car gets road-safety exemptions", *The Register*, 4 July 2011; retrieved 11 July 2011.

[4] Wolf-Heinrich Hucho, Chapter 1 - Introduction to automobile aerodynamics, *Aerodynamics of Road Vehicles, 1987, Pages 1-46.*

[5] McBeath, S., *Competition Car Aerodynamics: A Practical Handbook*, Haynes Publishing, ISBN 978-1844252305, 2006.

[6] WU Dawei, WU Zheng, ZhANG Landing, HU Jizhong, Safety and Airworthiness Design of Ultra-Light and Very Light Amphibious Aircrafts, The 2nd International Symposium on Aircraft Airworthiness (ISAA 2011), Procedia Engineering 17 (2011) 212 – 225.

[7] Haines, Thomas B. (19 March 2009). "First roadable airplane takes flight". *Aircraft Owners and Pilots Association (AOPA).* Retrieved 2009-03-19.

[8] Chien-Hsiung Tsai, Lung-Ming Fu, Chang-Hsien Tai, Yen-Loung Huang, Jik-Chang Leong, Computational aero-acoustic analysis of a passenger car with a rear spoiler, *Applied Mathematical Modelling, Volume 33, Issue 9, September 2009, Pages 3661-3673.*

[9] K.R. Cooper, Bluff-body aerodynamics as applied to vehicles, *Journal of Wind Engineering and Industrial Aerodynamics, Volume 49, Issues 1–3, December 1993, Pages 1-21.*

[10] Durbin, Dee-Ann (2012-04-02). "Flying car gets closer to reality with test flight". *boston.com.* Associated Press. Retrieved April 20, 2012.

[11] Mone, Gregory (2008-10). "The Driving Airplane Gets Real". *Popular Science*: pp. 42–48. Retrieved 2009-03-20.

[12] Phillips, Matt (March 18, 2009). "Flying Car Takes First Flight". *The Middle Seat Terminal* (The Wall St. Journal). Retrieved 2009-03-19.

[13] Welsh, Jonathan (April 5, 2012). "Flying Car Maker Offers 'Show Special' Discount". *Driver's Seat.* Wall St. Journal. Retrieved 2012-07-30.

[14] Terrafugia's Transition street-legal airplane continues flight and drive testing". Terrafugia. Retrieved 30 July 2012.

[15] Ryan, David L. (2009-03-18). "'Flying car' at the Museum of Science". Boston.com (The Boston Globe). Retrieved 2009-03-28.

CFD Modeling of Insect Flight at Low Reynolds Number

Wu Di, Yeo Khoon Seng, Lim Tee Tai

***Abstract*—**The typical insects employ a flapping-wing mode of flight. The numerical simulations on free flight of a model fruit fly (Re=143) including hovering and are presented in this paper. Unsteady aerodynamics around a flapping insect is studied by solving the three-dimensional Newtonian dynamics of the flyer coupled with Navier-Stokes equations. A hybrid-grid scheme (Generalized Finite Difference Method) that combines great geometry flexibility and accuracy of moving boundary definition is employed for obtaining flow dynamics. The results show good points of agreement and consistency with the outcomes and analyses of other researchers, which validate the computational model and demonstrate the feasibility of this computational approach on analyzing fluid phenomena in insect flight. The present modeling approach also offers a promising route of investigation that could complement as well as overcome some of the limitations of physical experiments in the study of free flight aerodynamics of insects. The results are potentially useful for the design of biomimetic flapping-wing flyers.

***Keywords*—**Free hovering flight, flapping wings, fruit fly, insect aerodynamics, leading edge vortex (LEV), computational fluid dynamics (CFD), Navier-Stokes equations (N-S), fluid structure interaction (FSI), generalized finite-difference method (GFD).

I. Introduction

With the advent in the progressive research for the applications and fabrication of autonomous micro aerial vehicles (MAV) or unmanned air vehicles (UAV), flapping mode flight is largely and widely studied in the past few years. With a wing spanwise of less than 10cm and a flight speed a few meters per second, MAVs experience the same low Reynolds number ($10^0 - 10^5$) flight conditions as their biological counterparts. In this flow regime, flapping wings type flight gains more efficiency, maneuverability and a wide range of speed over the rigid fixed wings mode, which might be the reason of its presence in nature among millions of species of birds and insects. However, researchers have long realized the "quasi-steady" aerodynamics assumption is unable to capture the physical phenomena or forces generation in flapping flight properly at this scale where the unsteady state flow mechanism dominates. In the past few decades, much work has been done to study the aerodynamics and energetics of insect flight and considerable progress has been made in these areas. The studies of Ellington [1]-[6], Dickinson [7], Ellington et al. [8], Dickinson et al. [9], Sane and Dickinson [10], Birch and Dickinson [11] and others revealed unsteady fluid mechanism in experimental aspects. Computer-based studies have also begun to play a bigger role in recent years to help unravel the intricacies of insect flight, most notably through the works of Liu et al. [12], Wang [13], Ramamurti and Sandberg [14], Sun and Tang [15] and Aono et al. [16]. These efforts have been successful in identifying and clarifying the high lift mechanisms available to flapping-wing insects to overcome weight.

Few works have focused on numerically studying the unsteady flows around 3D free flapping flyer, largely owing to difficulties in modeling the complicated geometry and in simulating the 3D movements of wings. Additionally, the kinematics and kinetics are inferred from observations of subjects, whose behavior it can be very difficult to control. As consequences, the measurements obtained from tethered insect may not truly reflect conditions in free flight. In this paper, we present a computational study on the low Reynolds number flight of a model fruit fly. The numerical model integrates the computational fluid dynamics of the flow with the three-dimensional Newtonian dynamics of the flyer and the governing Navier-Stokes equations. In contrast to the other computational studies, the wing kinematics employed in this work is not acquired from published experimental data. Instead, a basic set of sinusoidal wing kinematics are provided and allowed to evolve in time to achieve designated hovering flight with small deviations. Stroke plane adjustment and mean positional angle shifting are employed to regulate airspeed and stabilize the body pitch angle, respectively. Visualization of flow field around the model fruit fly in flight has also been obtained.

II. Materials and Methods

A. Morphological and Kinematic Model

For the present work, we have adopted the morphological data on fruit fly *Drosophila melanogaster* from Fry et al. [17] for our computer model of a low-Re flapping wing flyer. The geometric model of the flyer itself was constructed from photographic images of the insect. The model flyer, referred to here as the *model insect*, has a body length of 2.78 mm and wing length $R = 2.39mm$. The wings are assumed to be rigid and of uniform thickness equal to 2% of wing length. Considering that the aerodynamic influence of the wing root is likely to be negligible, we did not model the wing base for simplicity, but instead curtailed the wings at a short distance

Wu Di, Research Fellow, is with the Department of Mechanical Engineering, National University of Singapore, 9 Engineering Drive 1, Singapore 117576 (e-mail: mpewudi@nus.edu.sg).

Yeo Khoon Seng. Associate Professor, is with the Department of Mechanical Engineering, National University of Singapore, 9 Engineering Drive 1, Singapore 117576(phone: +65-6516-2246; fax: +65-67791459; e-mail: mpeyeoks@nus.edu.sg).

Lim Tee Tai, Professor, is with the Department of Mechanical Engineering, National University of Singapore, 9 Engineering Drive 1, Singapore 117576, (e-mail: mpelimtt@nus.edu.sg).

from the body.

According to measured data in [17], the weight of insect is $W = 0.96mg$, wing mean chord $\bar{c} = 0.874mm$, stroke angle $\Phi = 2.44rad.$ (140°) and average wing flapping frequency $f = 218Hz$. The Reynolds number in the hovering flight can be defined as

$$\mathrm{Re} = \frac{\bar{c}\cdot U_{ref}}{\nu} = \frac{R\cdot 2\Phi\cdot f\cdot \bar{c}}{\nu} = \frac{4R^2\cdot\Phi\cdot f}{\nu\cdot AR} \tag{1}$$

where ν is the kinematic viscosity of air ($1.5\times10^{-5}m^2s^{-1}$), U_{ref} the average flapping velocity of the wing tip and aspect ratio $AR = (2R)^2/S$ with the surface area of a wing pair $S = 2R\bar{c}$. The Reynolds number Re thus works out to be nominally around 143.

B. Fluid Dynamics Modeling

The dynamics of the fluid is governed by the incompressible non-dimensional Navier-Stokes (NS) equations, given in the arbitrary Lagrangian-Eulerian (ALE) form:

$$\nabla\cdot u = 0 \tag{2}$$

$$\partial_t u = -(u-u^g)\cdot\nabla u + \frac{1}{\mathrm{Re}}\Delta u - \nabla p \tag{3}$$

where $u(x,t)$ and $p(x,t)$ ($x \in \Omega(t)$)represent the velocity and pressure fields of the fluid domain $\Omega(t)$, respectively. The u^g is the convection velocity of the computational node, which is equal to zero for stationary nodes. The above equations are solved on a hybrid Cartesian cum meshfree grid system, wherein immersed bodies/boundaries and their near fluid neighborhoods are discretized by meshfree nodes. The meshfree nodes convect with the motion of bodies/boundaries. Fig. 1 shows the nodal system on the model insect generated using an available finite-element grid generator. The flyer is located in a cubic domain with boundaries set at a distance of about 8 wing-lengths away flyer. The cubic domain is discretized by a 221^3 Cartesian grid system (See Fig. 2). The grid is non-uniform with coarse grid on the outside and fine grid near the center where the model insect is located to give good resolution to the immediate flow around the flyer. The final grid has been selected through trials and comparison with experiments to ensure well-converged numerical performance.

Solution by the present hybrid-grid scheme utilizes a projection-based method, which has been described by Chew et al. [18] and Yu et al. [19], and will not be elaborated here. The hybrid grid combines great geometry flexibility and accuracy of boundary definition with the computational efficiency of a Cartesian background grid used in bulk of the flow domain. The cloud of meshfree nodes employed around the model also gives good resolution to the fluid boundary layer. In the present case, the meshfree nodes constitute around 1% of the total nodal population. The Cartesian background grid/nodes are stationary in space in most applications, but they can also be convected with the overall motion of the model flyer if necessary. The latter is particularly useful in problems where the flyer is allowed to traverse freely through space, such as in forward or backward flight.

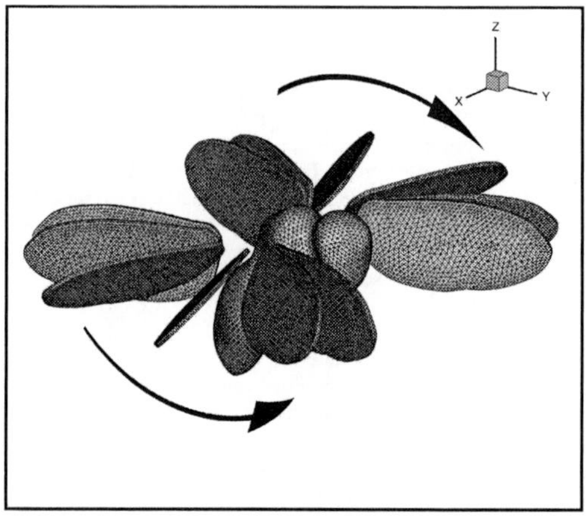

Fig. 1 Discretized fruit fly model in downstroke flapping

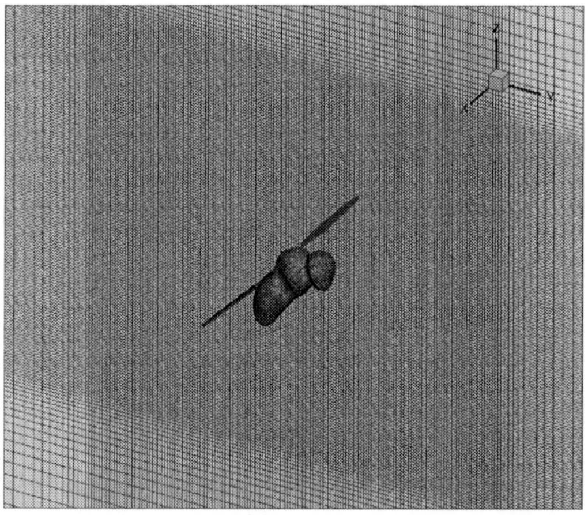

Fig. 2 Comparison between the messless points and its background Cartesian mesh

C. Fluid-Body Interaction

The flapping wing action of the model flyer drives the flow, whose dynamics has been stated above. The reaction force of the fluid in turn motivates the motion of the flyer in accordance with Newton's laws. Let $V_C(t)$ and $\omega(t)$ denote the linear and angular velocities of the flyer of mass M at its center of mass C, respectively. The governing dynamical equations for the flyer are then given by:

$$\left.\begin{aligned} \frac{dX_C(t)}{dt} &= V_C(t) \\ \frac{d\Theta(t)}{dt} &= [K(t)]\cdot\omega(t) \\ \frac{d(M\cdot V_C(t))}{dt} &= -Mg\cdot i_z + \int_{\Gamma(t)} [\sigma]\cdot n\, d\Gamma \\ \frac{d([I(t)]\cdot\omega(t))}{dt} &= \int_{\Gamma(t)} (x-X_C(t))\times([\sigma]\cdot n)\, d\Gamma \end{aligned}\right\} \qquad (4)$$

where $\Theta(t)$ is a rotation angle vector and $[K(t)]$ a transformation matrix. $[I(t)]$ is the inertia tensor of the flyer, while $[\sigma]$ is the Newtonian fluid stress tensor. The velocity of nodes on the moving body and wings of the model flyer is given by

$$u(x,t) = V_C(t) + V_{x/C}(t) + \omega(t)\times(x - X_C(t)) \qquad (5)$$

where $V_{x/C}(t)$ represents the velocity of nodes relative to the body frame of flyer, which is non-zero for nodes of the wings.

As can be seen, both equations of fluid (2), (3) and model flyer (4), (5) require information about the surface (body and wing) configuration $\Gamma(t)$ of the flyer. The time-dependent surface configuration $\Gamma(t, X_C(t), \Theta(t))$ comprises information about the current disposition of the flyer in its body frame, which is known, as well as information on the current spatial location $X_C(t)$ and orientation $\Theta(t)$ of the flyer, which are unknown. The solution of the fluid-body interaction problem at time *t* thus involves solving the coupled equations of the fluid and the flyer, given above, while simultaneously iterating on $X_C(t)$ and $\Theta(t)$ to determine the configuration $\Gamma(t, X_C(t), \Theta(t))$ that satisfies all the governing equations and conditions (2) to (5). This is essentially a fixed-point iteration problem, which could be symbolically represented by the following process equation:

$$\Gamma(t) = S(F(\Gamma(t))), \qquad (6)$$

where *F* symbolically denotes the flow solver, which computes the force/moment loading on the currently estimated configuration $\Gamma(t)$, and *S* is the body dynamic solver, which computes an updated configuration $\Gamma(t)$ of the flyer. This process of estimation and update is repeated until the change between consecutive estimates of $\Gamma(t)$ is less than a prescribed small real number ε, the convergence criterion. The free flight of the model insect is then described by the solution of $\Gamma(t)$ as a function of time. The above methodology has been described in some detail and validated in fluid-body interaction study by Yu et al. [19].

III. Results and Discussion

Fig. 3 shows the free normal hovering dynamics of the model fruit fly over a period of 40 wing cycles. The *y*-component of displacement fluctuation is the largest (Fig. 3A), while pitching is the largest component of rotational fluctuation (Fig. 3B), both within individual wing cycles and over the course of hovering simulation. Over the course of the 40 wing cycles, the body deviates<4% of a wing length from the designed hovering position (Fig. 3A), while the pitching angle of the body $\chi(t)$ oscillates within a narrow band of $< 3°$ after the 10th cycle (Fig. 3B). In fact, both displacement and pitching oscillations decrease with increasing wing cycles, showing a gradual convergence towards a sustained quasi-steady hovering state. The larger body fluctuations registered at the beginning is due to the flyer being initiated into flight at time $t=0$ from a rest state.

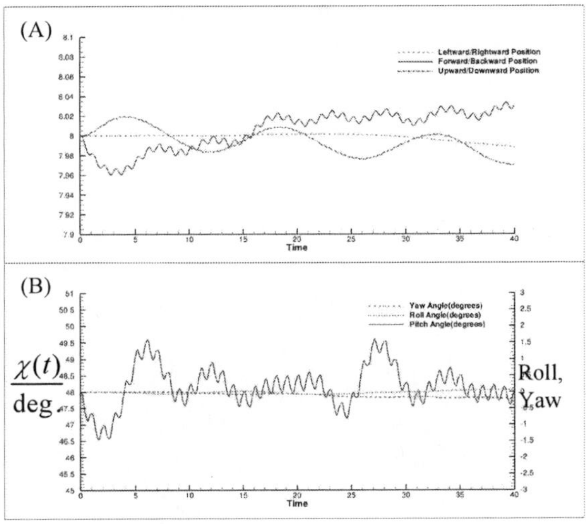

Fig. 3 (A) Translational fluctuation of the center of mass of flyer about the designated mean position. (B) Fluctuations of body orientation – mean body pitch angle $\bar{\chi} \approx 48°$ from the horizontal

Unsteady fluid flow over flapping wings is governed by the dynamics of strong vortices generated at the edges of the wings. The vortical flow over the wings of the model fruit fly comprises a leading-edge vortex (LEV) over the straight anterior margin of the wing, a wing-tip vortex (WTV) and a trailing-edge vortex (TEV). The term WTV is used here to distinguish the strong vortex that extends contiguously from the LEV over the curved section of leading edge in the distal 40% of the wing (see Fig. 4).

Fig.5 shows the vortex generation and shedding processes in hovering flight. The associated velocity fields in a lateral plane cutting across the vortex system at the wing root are also given. The bulk of vortical flow is shed via the WTV and TEV from each wing in a fairly continuous fashion during a stroke in an arc about the body. These vortices generated are roughly symmetric and of equal magnitude in the upstroke and downstroke during normal hovering, where the stroke plane is nearly horizontal. One may also observe the jets of flow passing between the WTV and the TEV. A stacked sequence of these vortices about each wing drives the flow below the wing.

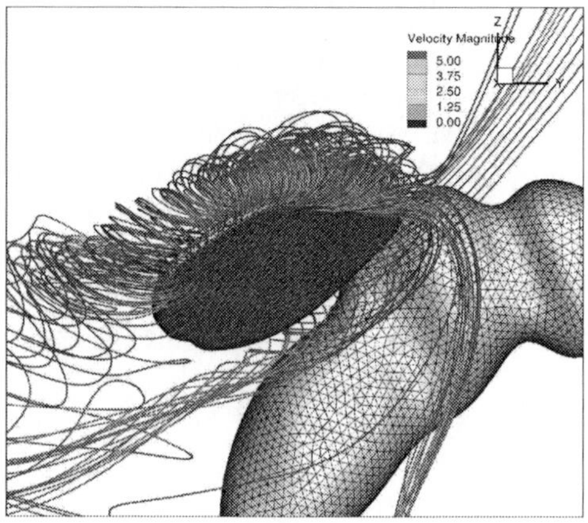

Fig. 4 Three-dimensional stream traces around the leading edge and wing tip regions

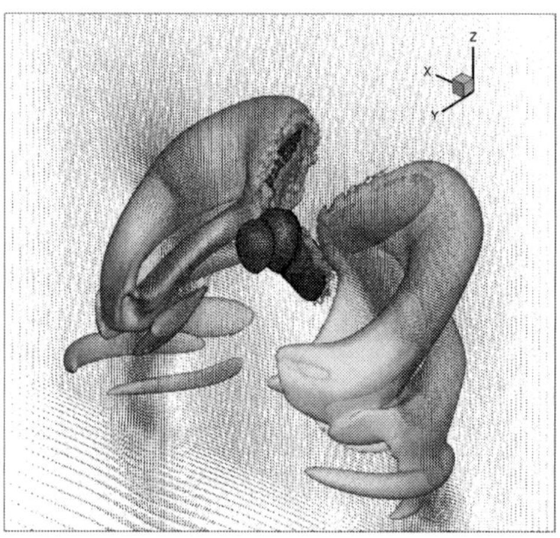

Fig. 5 Iso-vorticity surfaces around the model fruit fly during normal hovering. Also given is the velocity field in a lateral plane cutting across the vortex system

IV. Conclusions

This paper is concerned with a numerical study on the free hovering of a low Reynolds number flapping-wing flyer, modeled after the fruit fly. The unsteady flow is modeled by the incompressible Navier-Stokes equations, while the dynamics of the flyer is governed by Newton's laws of free body in a gravity field. The deviation of center of mass of the insect from the designated hovering position is typically less than 4% of one wing length while the orientation or pitching angle of the body varies within a band of less than 3 degrees for hovering flight. Flow field study shows that the flow jets induced by the wings are broadly similar to those in published results for normal hovering. The present modeling approach offers a promising route of investigation that could complement as well as overcome some of the limitations of experiments in the area of free flight aerodynamics of insects.

Acknowledgment

The generous financial support for the present work under the research grant TDSI/09-007/1A from the TDSI (National University of Singapore) is gratefully acknowledged.

References

[1] Ellington, C. P.. The aerodynamics of hovering insect flight. I. The quasi-steady analysis. *Phil. Trans. R. Soc. Lond. B* 1984; 305: 1-15.

[2] Ellington, C. P.. The aerodynamics of hovering insect flight. II.Morphological parameters. *Phil. Trans. R. Soc. Lond. B* 1984; 305: 17-40.

[3] Ellington, C. P.. The aerodynamics of hovering insect flight. III. Kinematics. *Phil. Trans. R. Soc. Lond. B* 1984; 305: 41-78.

[4] Ellington, C. P.. The aerodynamics of hovering insect flight. IV. Aerodynamic mechanisms. *Phil. Trans. R. Soc. Lond. B* 1984; 305: 79-113.

[5] Ellington, C. P.. The aerodynamics of hovering insect flight. V. A vortex theory. *Phil. Trans. R. Soc. Lond. B* 1984; 305: 115-144.

[6] Ellington, C. P.. The aerodynamics of hovering insect flight. VI. Lift and power requirements. *Phil. Trans. R. Soc. Lond. B* 1984; 305: 145-181.

[7] Dickinson, M. H.. The effects of wing rotation on unsteady aerodynamic performance at low Reynolds numbers. *J. Exp. Biol.* 1994; 192: 179-206.

[8] Ellington C. P., van den Berg, C., Willmott A. P. and Thomas A. L. R.. Leading-edge vortices in insect flight. *Nature* 1996; 384: 626-630.

[9] Dickinson, M. H., Lehmann, F.-O. and Sane, S. P.. Wing rotation and the aerodynamic basis of insect flight. *Science* 1999; 284: 1954–1960.

[10] Sane S.P. and Dickinson M.H.. The control of flight force by a flapping wing: lift and drag production. *J. Exp. Biol.* 2001; 204: 2607-2626.

[11] Birch J.M. and Dickinson M.H.. The influence of wing-wake interactions on the production of aerodynamic forces in flapping flight. *J. Exp. Biol.* 2003; 206: 2257-2272.

[12] Liu, H., Ellington, C. P., Kawachi, K., Van den Berg, C. and Willmott, A. P.. A computational fluid dynamic study of hawkmoth hovering. *J. Exp. Biol.* 1998; 201: 461-477.

[13] Wang, Z. J..Two dimensional mechanism for insect hovering. *Phys. Rev. Lett.* 2000; 85: 2216-2219.

[14] Ramamurti, R. and Sandberg, W. C.. A three-dimensional computational study of the aerodynamic mechanisms of insect flight. *J. Exp.Biol.* 2002;205: 1507-1518.

[15] Sun, M. and Tang, J.. Unsteady aerodynamic force generation by a model fruit fly wing in flapping motion. *J. Exp. Biol.* 2002; 205: 55-70.

[16] Aono H., Liang F. and Liu H.. Near- and far-field aerodynamics in insect hovering flight: an integrated computational study. *J. Exp. Biol.* 2008; 211: 239-257.

[17] Fry S. N., Sayaman R. and Dickinson, M. H.. The aerodynamics of hovering flight in Drosophila. *J. Exp. Biol.* 2005; 208: 2303-2318.

[18] Chew C.S., Yeo K.S. and Shu C.. A generalized finite-difference (GFD) ALE scheme for incompressible flows around moving solid bodies on hybrid meshfree-Cartesian grids. *J. Comput. Phys.* 2006; 218: 510-548.

[19] Yu, P., Yeo, K.S., ShyamSundar, D. and Ang, S.J.. A three-dimensional hybrid meshfree-Cartesian scheme for fluid–body interaction.*Int. J. Numer. Meth. Engng.* 2011; 88: 385–408.

AUTHOR INDEX